CHAOS AND LIFE

COLUMBIA UNIVERSITY PRESS / NEW YORK

CHAOS AND LIFE

COMPLEXITY AND ORDER IN EVOLUTION AND THOUGHT

RICHARD J. BIRD

COLUMBIA UNIVERSITY PRESS
Publishers Since 1893
New York Chichester, West Sussex

Library of Congress Cataloging-in-Publication Data
Bird, Richard J.
Chaos and life : complexity and order in evolution and thought / Richard J. Bird
p. cm.
Includes bibliographical references and index.
ISBN 0–231–12662–X
1. Biology—Philosophy. 2. Evolution (Biology)—Philosophy. 3. Chaotic behavior in systems. I. Title.

QH331.B525 2003
570'.1—dc21
2003051637

Columbia University Press books are printed
on permanent and durable acid-free paper.
Printed in the United States of America
Designed by Lisa Hamm
c 10 9 8 7 6 5 4 3 2

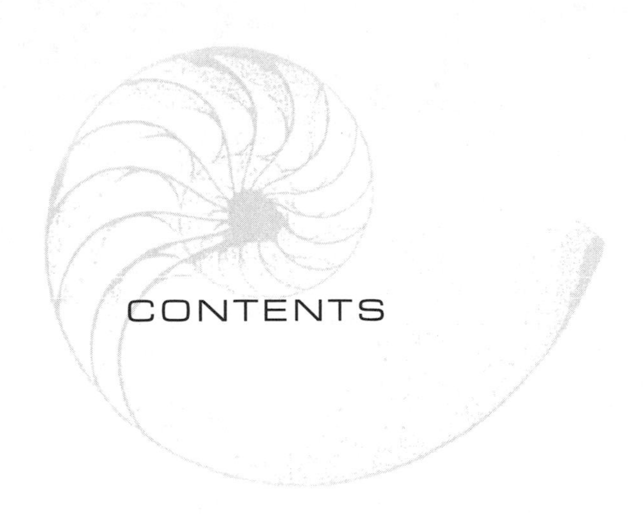

CONTENTS

PREFACE

This is a book about chaos and life, and it mostly tells its own story. If there is a subtext it is that the scientific enterprise, which has been the inspiration of our age, is now flagging in that role. I try to explore some of the reasons behind the failure of science to engage the emotions and loyalties of the general population it serves. In my view this is because science has, at least since the Enlightenment, neglected those aspects of reality that it cannot comfortably incorporate into its associated edifice of belief. This has come to be summed up in the faith, especially entrenched in biology, that events are essentially random and that from this random set of events some are selected because they are better adapted than others to their surroundings. This belief, which is an axiom of Darwinist explanations, renders essentially meaningless those coincidences and apparently significant events that fill the everyday lives of men and women. It is an extraordinary fact that, in a scientific age, people turn to astrologers, mediums, cultists, and every kind of irrational practitioner rather than to science to meet their spiritual needs. This is a gap that science must fill, and fill urgently. This book is in part an attempt to show how this might be done.

AN EMERGING WORLDVIEW

The central theme of this book can be briefly stated: the world is a series of iterative systems; from iterative processes sequences emerge that carry information; our understanding of the world depends on our interpretation of sequence; events in the world occur not randomly, but in an organized way, which means that nature holds many surprises for us as its large- and small-scale organization becomes better known.

Such ideas form the basis of a newly emerging worldview. As they become more widely understood I believe they will have a great impact on our outlook on the world and indeed on the world itself. Rather than the old Newtonian picture, a new worldview is arising: one of nature as an organic rather than an atomic process and the organization of events as sequential rather than random. As a consequence, we will begin to understand nature not as something separate from ourselves, but as the matrix in which life was set from the inception of the universe. We will come to see other living things as more closely related to ourselves than we have so far, because we are shaped by processes of the same kind, and we will better understand the inbuilt paradoxes and conflicts within our nature that result from them.

ACKNOWLEDGMENTS

Schopenhauer said that a friend in conversation was the midwife at the birth of a thought. If so then a number of my friends have certainly earned certificates in midwifery. Among them Richard Kenyon, Victor Serebriakoff, Robin Smith, Sherrie Reynolds, all of my family, Fred Abraham, Malcolm Weller, Elliott Middleton, and Boris Sømød were called upon to facilitate a delivery, sometimes in the middle of (their) night. Conversations not followed immediately by a birth with Sue Aylwin, Ronnie Cunliffe, John and Irmgarde Horsley, Jouko Seppänen, Adrian Spooner, and Alan Slater led more often than not to a pregnancy. I would also like to thank my colleagues at the University of Northumbria who allowed me to take sabbatical leave during a crucial point in the writing of the book. Among them, Chris Dracup, Pam Briggs, Delia Wakelin, and John Newton particularly helped me, and Jeremy Atkinson's encouragement was greatly appreciated. The students of my "Chaos and Fractals" courses over several years have contributed hugely to my understanding.

Of those outside the university, the late Sir Fred Hoyle had the greatest impact on my concept of mathematical morphology, and I am indebted to Rikki Brown and Chandler Davis for comments on the ideas expressed in chapter 9 and to the late Nick Whitfield for corrections and encouragement in chapter 10. Michael Richardson kindly supplied the pictures by Haeckel in chapter 7. Brian J. Ford gave me much encouragement with chapter 2, part of which appeared in the *Biologist* in 1996. Without the influence of Victor Serebriakoff's books *Brain* and *The Future of Intelligence* I am sure chapter 5 would never have been written. Jack Cohen's wise advice was much appreciated, as was the help of Margaret Comaskey, who improved many of my infelicities. I would like to say a particular "thank-you" to all the officers and members of the

Society for Chaos Theory in Psychology and the Life Sciences, for creating a forum where some of the most vital ideas of our time can be discussed.

I want to make a special mention of the debt that I owe to Robin Robertson, author, mathematician, and magician, whose explication of the work of George Spencer-Brown has helped me so much. His constant encouragement and detailed assistance were indispensable and without them this book might never have been completed.

The gnomic and fragmentary sayings of Heraclitus, some of which epigraph the text, have haunted me since I first encountered G. S. Kirk's translations and commentary and I wish I could have talked to them both.

Dick Bird
Newcastle 2003

CHAOS AND LIFE

PROLOG: THE DAWN OF MAN?

About a hundred thousand years ago, perhaps as many as a hundred and fifty thousand or as little as fifty thousand, the first man was born. In the eyes of his parents he must have been an ugly baby, an outcast from the brood and from the tribe from the moment of his birth. He was lucky to survive, for monstrous births were not generally suffered to live in that time and many more like him would have perished, and would in other times and places perish, by accident or design, and without further consequence. He was different from his tribe: his features to them seemed curiously unnatural: angular, sharp, and distorted. His gracile frame seemed spindly and ill adapted to survive in the rigorous climate, which could vary every few years, alternating between humid–hot and freezing–dry conditions, interspersed with a temperate mediocrity hardly more favorable, since it ill suited the growth of the plants on which the tribes depended. To the society of the race he was a useless individual—an idle jack—spending much time in seemingly depressed, introverted contemplation; staring into space; fiddling with pieces of stone and slate; or making marks in the earth with a stick. Perhaps his social experience was the origin of the story of the Ugly Duckling. The first Outsider, he must have been familiar with loneliness in a time when loneliness was a difficult condition to achieve and an undesirable one.

Let us call him Adam.

Adam sat for long periods between gathering plants or catching animals, when he was not shivering too much to think or too exhausted by the sun's heat. And in those periods when he was able to reflect, he contemplated the great mystery that was his existence in the world. He could speak in a way—

the gift of tongues—that the tribe could not, but he had no one to speak to. He made peculiar guttural and explosive sounds, and he seemed to gesture, in apparent madness, at the beasts and plants and rocks, and even at the sky, as he made these strange noises. This was the birth of language, and he was naming the things around him.

ITERATION AND SEQUENCE

It scatters and gathers, it comes together and flows away, approaches and departs.[1] (HERACLITUS, FRAGMENT 91)

ITERATION

Life, it is often said, is just one damned thing after another. It would be hard to think of a better description of iteration. Iteration is a ubiquitous phenomenon and the operations of the world are for the most part, if not entirely, iterative. Rivers flow day after day in almost (but not quite) the same channels. Winds blow in apparently repetitive seasonal patterns; planets rotate, revolve, and wobble in their orbits; some clocks have hands, which rotate too. Living things reproduce themselves from generation to generation in an iterative way and they grow by the division of their cells, which is also iterative. In our daily lives too, we iterate, by going to work, going shopping, washing up, and finally, when we fall exhausted into bed, by counting sheep.

Iteration is not simply repetition; it is a creative process. The ancient Greek philosopher Heraclitus said, "Upon those who step into the same rivers, different and again different waters flow." (He has also been interpreted as saying that "all things change," although he did not quite say that.) When a process iterates, it is performed over and over again, but each time with alterations and modifications that may be slight but are nevertheless productive. After one rotation of its axis the earth is not the same as it was before: the planet has moved on in its orbit round the sun and the name of the day has changed. A clock ticks and time has taken another step forward, its petty pace summing up in time to a majestic advance.

It might be as well to say clearly at the outset what I mean by iteration. Iteration is more than simple repetition. If something is done several times, if the same operation is performed repeatedly with no connection between one cycle and the next, I will call that simple repetition. To qualify as iteration a process must have two further steps. First, there must be an output from the operation that is used as an input by the next iterative cycle. Second, that output must be treated in a consistent way at each cycle.

For example, if you lay many bricks, you are repeating the operation of bricklaying. There does not have to be a necessary or consistent connection between one operation and the next; it all depends on how the bricks are laid, so the laying of bricks is, in general, a repetitive operation. However, when the bricks are laid one on top of another in a pile, something more than this is happening. The top of each brick dictates the position of the bottom of the next brick, and the distance between successive bricks is always the same. Building done in this way can be said to be an iterative process and it will produce some kind of construction.

Iteration involves time; it is intimately bound up with time and in a sense it defines time. Time is always measured by some device that iterates: a swinging pendulum, a vibrating crystal, the dripping of water, the rotation of the earth; all these are iterative systems. The duration of the iterative cycle of each particular system is its unit of time. For example, a pendulum swings in units of seconds, the crystal oscillates in microseconds, though in each case the units may be accumulated to form larger units such as seconds, minutes, or hours. Of course, all these kinds of measures are relative. We measure time in terms of the occurrence of events; there is no other way to measure it. For most everyday purposes we express time in units of fairly short duration, but for other purposes we measure time in different, much longer units. In astronomy, when considering such things as the movements of the earth's axis, we might want to measure time in millennia, or for evolutionary purposes the lifetime of a species may be measured in millions of years. However, there is no absolute measure of time that stands outside all these iterative processes and against which they can be judged. Some people may disagree with this; they may feel that there is some absolute standard of time—a smoothly running time that fits the differential equations used by physicists and is not composed of lumpy events forever succeeding one another. But no way of measuring such a "smooth" time has ever been found. All time as we know it is measured, and in a sense produced, by processes that are iterative, lumpy, and discontinuous.

The importance of iteration can be seen most clearly in a science that many now consider fundamental to our understanding of the world. This general approach is called dynamical systems theory, because it takes its inspiration from the study of fluid dynamics. Two of its aspects, which have become widely known in the last few years, are chaos and fractals, or more accurately, the study of chaotic systems and of fractal forms.

CHAOTIC SYSTEMS

Far from being random, as common usage implies, chaotic systems follow strict laws; that is why the functioning of such systems is called *deterministic chaos.* Chaotic systems are those that are influenced by very small changes in their controlling factors, like the weather, economics, social and management structures, and the brain and central nervous system. What these disparate things have in common is that they are often delicately poised between one state and another and so can easily be pushed in an entirely new direction by even very small forces. It takes only an item of bad news to set stock markets tumbling, or to push a person into depression. In psychology such changes are made more striking because information requires almost no energy for transmission; it takes very little effort to convey news, good or bad. But in other chaotic systems too, the force required to influence their long-term future is vanishingly small. This characteristic of chaotic systems is called *sensitivity to initial conditions.*

The other important feature of chaos is that it contains, and in a sense produces, order. The traditional use of the word *chaos* signifies complete disorder, but the modern science of deterministic chaos has shown that there is a great deal of orderliness in the patterns of movement of chaotic systems. These patterns can be visualized as often-beautiful geometric forms called "strange attractors." These forms can sometimes be used to enable us to forecast what will happen in such a system. Thus new meaning is found in the old expression, now being revived: "order within chaos."

A good example of chaos is the magnetic pendulum sold as an executive toy. It has four magnets arranged in a square at the base and a pendulum that swings back and forth between them. Release the pendulum and note the magnets that it visits, and in what order. If the pendulum is released from the same position a second time, the pattern of movement may at first be the same but soon it will become completely different. In fact, the pattern of its movement is chaotic. No matter how much care is taken to start the pendulum in the same position, it will visit an entirely different set of points on the two occasions.[2] Chaotic systems are generated by iteration, though not all iteration leads to chaos. In order to produce chaos, the iteration has to be within what is called a nonlinear system.[3] Nor are all nonlinear systems chaotic: to become so they need to be pushed beyond a certain point, called a *bifurcation.* Before that point is reached they may behave in a quite orderly fashion.

Economic systems are often, though not always, chaotic (in the technical rather than the popular sense!); for example, the movements of the price of gold or international currencies. For a long time people discussed whether markets were deterministic or random. A third view now seems more likely: that their operation can at times be chaotic—that is, they may be deterministic, but have a very complex pattern of changes that are sensitive to the surrounding conditions. Markets can often remain poised while everyone is waiting to see which way events will go. At some critical moment, buying can suddenly change to selling because of a small displacement of transactions in one or the other direction. If these movements are extreme, then a rapid downward slide or crash can occur.[4]

A fertile hunting ground for chaotic systems is the human brain, which in turn contains many subsystems, some of them chaotic also. Sensitivity to initial conditions predominates in our behavior. Everyone knows how the course of the day can often depend on some small event—a letter bringing good or bad news, a passing remark to someone on the way to work—that seems to set the pattern of the day. Thereafter other events seem to follow the pattern and establish what we call a "good day" or a "bad day." Probably what happens is that the sleep-refreshed brain is in a poised state (it is a system in a state of high-dimensional chaos): the first event to gain access to the emotional control centers pushes the system slightly one way or the other. In most cases the mood then tends to swing in a downward or upward direction for a while (sensitivity to initial conditions) until we regain the poise we started off with. In pathological cases an oscillating cycle may establish itself over a period of weeks or months and become manic-depressive illness, while in normal people the same process produces only the sort of much more rapidly changing moods that make us feel "up" or "down." Normal mood swings are chaotic, while manic depression is much more regular.

The first-studied and in some ways prototypical chaotic system is the weather. It has been asserted that there is a *butterfly effect* in the weather; that there are periods during which the masses of moving air are so delicately poised that a force as slight as the fluttering of a butterfly's wing would be sufficient to push the whole system one way or another and trigger changes culminating in a hurricane on the other side of the world. Although this has been criticized as a naïve view, chaotic models of weather do certainly imply it. The weather may also demonstrate another feature of chaos—the different amounts of variation that occur between one part of the chaotic cycle and another. From earliest times people have wanted to predict, or even better to control, the weather, but only recently has it been realized that this is harder to do at some times than at others because of changes in the variability of the system. At some points in the regime, conditions are quite stable and difficult to change; at others the butterfly effect comes into play. At these points, according to chaos theory, the smallest forces are sufficient to change events, and it becomes hard to foresee them, although paradoxically, at such a time they can easily be controlled; at other times they are more predictable and a butterfly-Canute trying to turn back the weather would beat its wings in vain.

Figure 1.1 shows a strange attractor that describes the weather. It is called the Lorenz attractor. It looks orderly though complex. In fact, the path of the attractor is one long, continuous track that theoretically never intersects itself and in time would visit every part of its domain, filling all the available space. Of course, there are great differences between the weather and a biological organism, and correspondingly different approaches to the study of chaos in biology from that taken in, say, meteorology or physics.

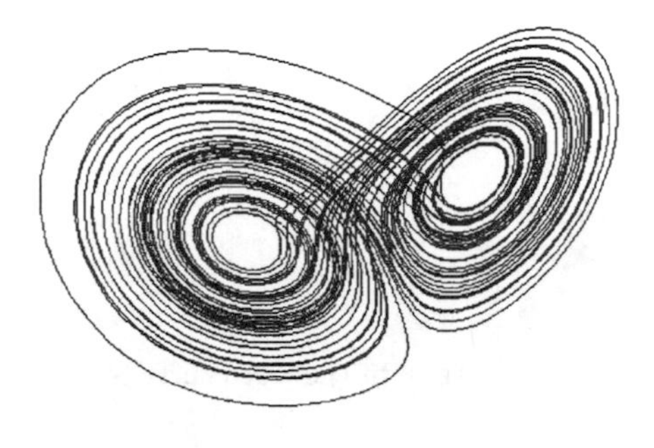

FIGURE 1.1 The Lorenz attractor.

FRACTAL FORM

The second process involving iteration that has become important in recent years is the production of shapes called fractals. A regular fractal is a curve in which a section is replaced by a new and more complex shape, whose sections are then replaced in their turn. At each stage the size of the curve section being replaced gets smaller and smaller. This gradually yields very detailed shapes indeed; figure 1.2 shows a complex curve made in this way called the Koch curve. (This looks a bit like a snowflake, and a snowflake is also a real-world fractal.)

The processes in the formation of this curve are very simple, although the resulting shape is very far from being simple, as shown in figure 1.3. The starting point is a straight line (a) whose middle section is replaced by two new lines (b). This process is continued (c), and after only a few steps the line has taken on a very complex appearance (d).

FIGURE 1.2 The Koch curve.

One remarkable feature of fractals is that they compress a tremendous amount into a small area. It is not hard to see that the curve in figure 1.3 is getting longer every time another replacement occurs, because each time we are increasing the length of that section by 1/3. In fact, if we go on and on replacing sections, the Koch curve will become as long as we like to make it. In this way we can make a curve that is infinitely long although it is contained in a finite area. As Ariel says in *The Tempest,* "I could count myself king of infinite space and yet be bounded in a nutshell."

Figure 1.4 is a picture of another fractal, called the Hilbert curve, which is also formed by replacement. This curve has an even more remarkable property: if the process of replacement of segments is taken far enough, it fills up the area inside its boundary because it will eventually pass through any chosen point. However, it is still "only" a line.

We have already seen a space-filling curve in the context of the Lorenz strange attractor, so it may not be a surprise to discover that strange attractors are fractals, although not regular ones. This is one of the links between the processes creating fractals and those generating chaos.

FIGURE 1.3 Stages in the formation of the Koch curve.

The reason fractals have become so important is that they are not just a mathematical curiosity but are everywhere in the real world. Benoit Mandelbrot, who first introduced the concept of fractals in his book *The Fractal Geometry of Nature,*[5] asked the question "how long is the coastline of Britain?" You might think that this question could be answered easily, but it is not as easy as it appears. How is the measurement to be made? If you take a map of Britain and measure around it with a pair of dividers, you will get a series of measurements that can be used to work out a result. Suppose you use a setting of the dividers that corresponds to ten miles. This will produce an estimate of the length of the coast, but it is not a final one, for in using this interval a great deal of detail has been left out. If instead we use a smaller separation of the dividers, say one mile, then the distance will be greater because the path will now go in and out of more of the bays and peninsulas that were previously bypassed. If you continue to decrease the measuring interval, the detail of the

FIGURE 1.4 The Hilbert curve.

map you are using will be exhausted and a larger scale map will become necessary. Eventually, any map will become useless and you will have to resort to going around the coast in a motorboat or on a bicycle. But even this measurement will not give a true answer, for what about the distance around the tiny inlets? The next stage would be to measure the ins and outs of the rocks and this adds more distance; then the pebbles, and even the sand grains themselves. Theoretically, if not practically, this could go on forever, and the coastline can be made as long as you like. This thought-experiment reveals an important attribute of all fractals, regular or natural, which is *self-similarity*. The coastline looks very much the same from different heights and pilots say that landscape features repeat themselves at different altitudes. Self-similarity is particularly important in nature, and we will return to this topic in chapter 4.

But it is not just coastlines that are fractal. Fractals are found in a multitude of places in the inorganic inanimate and organic animate world. Plants and animals are also fractals, though of a slightly different kind, and a multiplicity of forms, from river basins to nerves in the heart, can be described by the mathematics of fractals.

These two areas, fractals and chaos, recur many times in this book. Both are dependent upon a repeated sequence of operations; yet beyond them lies the more fundamental operation of iteration itself. To obtain a fractal curve or a chaotic orbit a sequence must be iterated many times. It appears that this is one of the fundamental things about the world: those processes that are most basic to its structure involve iteration.

ITERATION AND TIME

Perhaps the most basic level at which iterative processes appear in the world is that of time. It is very difficult to say what time is because, although it is self-evidently there, it is in some way intangible. Attempts to give a definition of time almost always come to grief in problems of self-reference. If asked what time is, people will usually say something like "time is the succession of events"; but what is a succession? "something involving time"; so we are no further forward.[6]

A better way to approach the subject is to ask how time is measured; if we know how something is measured operationally, then we may begin to get some idea of what it is. Here we are on firmer ground, because time is measured by clocks, and different kinds of clocks tick at different rates. In a way corresponding to these differing rates we measure time according to different time scales, from the briefest of chemical reactions measured in femtoseconds[7] to the cycle of astronomical events measured in thousands of years. These time scales are different and cannot usefully be derived from one another: it would be as useless to express the time of rotation of the earth's axis in seconds as it would be to time a chemical reaction by means of a sundial.

We usually think of a clock as a physical thing, like an alarm clock or a wristwatch. But a clock is really a process embodied in a machine,[8] and the nature of that process, as we noted at the start, is iterative. A clock can be almost any process that repeats itself over and over again for an indefinite period. Water clocks drip at a steady pace; quartz crystals vibrate regularly. Indeed, it is almost impossible to think of a clock that does not depend on a repetitive cycle of events. The only example that comes to mind readily is a candle marked in hours. But here too there is iteration—the repeated burning of molecules of wax—so this too is an iterative process, although at first disguised. The use of radiocarbon dating is another, much longer scale clock that also appears to be like this. It seems to yield a smooth time scale but in fact does not: the decay of atoms of carbon-14 is repetitive, although on a large scale it gives the appearance of being continuous.

Mathematical iteration is at first sight something rather different from this sort of real-world iteration. In mathematics, to iterate means to perform the same operation, usually an arithmetic operation, repeatedly. The result of each cycle of the operation is used as the input to the next cycle. The results of the successive iterative cycles are called the *iterates*.[9]

It is tempting, then, to equate time and space in the real world with processes that can be described by iteration. In the real world it takes time to iterate; iteration thus implies time. Iteration as a real-world process also implies the space in which to iterate. This parallel between the abstract process of iteration and the real world of space and time gives some justification for the statement that the world can be completely described by iteration. This is a question to which I will return in the final chapter.

SEQUENCE

Iteration produces sequence. Think of a deck of cards. When you buy it the cards are usually arranged in order from the king of spades to the ace of clubs. This is a prearranged sequence and before you can play a game with the cards you have to rearrange it by thorough shuffling. There are various methods of shuffling cards. Sometimes people deal them into five or seven piles in between the more usual shuffles, where a handful of cards is taken and some repeatedly distributed on either side of the rest of the deck. Also common is the riffle-shuffle, where the deck is divided into two and flipped together so that the cards interleave alternately, or nearly so. Playing a game in itself shuffles cards into a new order, and so it goes on, with new sequences forming and dissolving throughout the life of the deck, like eddies in a river.

There are 52 cards and the number of ways they can be sequenced[10] (if you count circular arrangements as being equivalent to each other) is a number with 67 digits in it[11]—big enough to satisfy the most demanding appetite for variety. You would be very unlikely to see exactly the same disposition of cards twice in a hand you were dealt. There are frequent reports of games in which everyone has a "perfect" hand—one player holding all the clubs, another all the hearts, and so on. The probability of this is a much less astonishing number—about one in 600 billion. If these hands really do happen and are not just tall stories, it must be because the order in the cards must have been preserved from a previous deal. If you sit down with a gambler expecting that the cards will be distributed randomly, you will lose your money, and not necessarily because the deal is crooked. Shuffling is popularly supposed to remove order from the cards, but an exact interleaved riffle-shuffle (the Faro Shuffle, easily done by a conjuror) will restore a deck of 52 cards to its original sequence in just eight moves.[12] There is literally more to sequence than meets the eye.

The same applies to many other supposedly "random" arrangements of things. Consider a Rubik's cube. At first it looks like a mechanical impossibility, but if you take one apart,[13] you can see how ingeniously it works. The mechanical impossibility is equaled by the mathematical complexity. It is possible to twist each of the cube's faces through a whole number of right angles in either direction to produce a new configuration. The Rubik's cube is not as complex as a deck of cards: it has only about 10^{19} different arrangements. This is still enough to make it difficult to get the same arrangement twice running unless you are trying to; it would take 1.4 million years to run through all the arrangements if each took only a microsecond. Suppose you twist the faces in turn through ninety degrees, also rotating the whole cube through ninety degrees between each twist: it will rapidly assume a random appearance. However, the rule for producing this arrangement of faces is a very simple one and can be easily reversed in a few twists by reversing the operations. To someone who did not know how it had been done, it might look impossible to get the cube back to its original position in only a few moves. It would be necessary to apply a general method for restoring the cube, the most efficient of which takes about 85 moves. In fact, the cube is never more than 20 moves away from its starting point.[14]

The sequence of events is all-important. It is important to language; reversing the order of words in a sentence can contradict its original meaning or turn it from a statement into a question. In mathematics, the order of the propositions of a theorem is essential to its truth. In music, the sequence of notes creates a tune, and altering the sequence will change it into a different tune. In DNA, the sequence of the bases A, G, C, and T prescribes the form and function of the organism: the alteration of that sequence by substituting one base for another can wreak havoc in the unfortunate person born with a genetic defect. Our place in a queue is very important to us: the order in which we get treatment in a hospital emergency room or even a post office may make a big difference to our lives. We spend much of our time rearranging, sorting, and ordering objects, events, and people. The ancient Chinese art of arrangement, Feng Shui, prescribes rigid patterns for the placement and order of objects in a house and even for the position and orientation of the house itself. Many rational people think this is just a superstitious belief but, whether valid or not, the philosophy of Feng Shui is another example of the universal appreciation of the importance of sequence.

Examples of sequence are all around us. The sequence of atoms in a molecule is what makes the difference between one chemical substance and another. The sequence of keys on a keyboard is necessary for efficient typing; keys transposed to unexpected places lead to errors, as happens with the U.S. and U.K. computer keyboard layouts. The whole numbers (integers) are arranged in a fixed ascending sequence; letters in an alphabet are always written in the same order. The sequence in which vehicles move on a road can make an accident likely or unlikely. During the growth of an organism, events must unfold in the correct sequence for normal development. Books in libraries, words in dictionaries, articles in stores, moves in a game of chess, runners on a curtain rail, molecules of water, dancers in a ballroom are all arranged in a prescribed order. The leaves of plants, the taxiing of airplanes, the spinning of the roulette wheel, the "playing on the merry organ," and the "sweet singing in the choir" depend upon sequence, its creation and preservation. A different sequence would produce a different result, just as a different arrangement of the examples in this paragraph would have made a different impression on the mind. Obvious examples, one might say, but it would not be too much to say that every new property emerges from a new sequence, or arrangement, of a set of objects.

THE PERSISTENCE OF SEQUENCE

Another universal property of sequence is that once established, it is hard to destroy. The robustness of sequences is very noticeable at a personal level. Everyone must have had the experience of entering an amusement park, supermarket, or exhibition with some other people whom they don't particularly like the look of, and then finding it impossible to lose them. The same thing seems to happen with drivers on a highway who get "stuck to your tail." The persistence of sequence is used by stage magicians, who depend on the inadequacy of shuffling to destroy the sequence in a deck of cards.[15] It has been found to operate in mixtures of various kinds, where fluids are stirred in a tank or in a domestic food mixer; patterns have been shown to recur with every cycle of the mix.[16] A familiar experience to any cook is the lump of some ingredient that sticks to the side of the mixing bowl and goes around and around without being sliced up and mixed in. These stable objects can be found in systems large and small. A tornado is a stable vortex with a tremendous vitality for survival. On an even vaster scale, so is the Great Red Spot of Jupiter, apparently an enormous hurricane that has lasted for hundreds of years (possibly much longer) without disintegrating.

While sequence and order effects are now being studied in science for the first time, they have always played a large part in everyday life. I will mention just three examples to show how diverse are the areas in which they occur. First, consider the effects of sequence on food and drugs. A food taken by itself will have a particular effect, but if it is combined with another food it may have a quite different one. Some foods do not go well together; cheese and chocolate, for example, or almonds and mushrooms, do not make good combinations. However, the effects may be much more profound that this. Studies at the Free University of Brussels as far back as the nineteen-fifties showed that the dietetic effects of foods are not isolated but are quite marked if the foods are taken in combination with others. For example, protein and fat taken together have a different effect than when taken separately, and many more similar results have since been established.[17]

Every cook knows that the way in which food is mixed is critical for successful results. Especially critical are sauces, which are notorious for being very sensitive to how they are prepared. It is often difficult to pin down the factors involved. One is certainly temperature (which increases the available molecular sequences) but another may be the type of mixing applied. Grinding spices by hand is considered more effective than grinding them mechanically; stirring is essential for most cooking.

What happens with food is mirrored in drug manufacture and administration. Many drugs are synergistic in their action; for example, aspirin and codeine may be more effective as painkillers if given together than either by itself or both given separately. Aspirin, codeine, and caffeine taken together are a powerful combination and are most effective when combined in a mixture. Taking an aspirin, then a codeine tablet, and then drinking a cup of coffee one after the other is not as effective as taking a mixture of the three at one time. However, the effects extend further than this: one drug combination may not be as effective as another even with the same ingredients; the difference may be a subtle one and may depend not on the chemical content so much as on the mode of preparation. It is currently again in the news that the MMR triple vaccine (against measles, mumps, and rubella) given to children in Britain as a combined injection is alleged to have caused some cases of autism, while if the vaccines are given separately over a period of months this risk is supposed to be reduced.[18]

As a second example, consider the effects of order in catalytic chemical reactions. Catalytic processes are those in which chemicals react together but the catalyst itself is not changed. It has recently been found that the order of molecules in the catalytic reaction known as hydrogenation controls its efficiency.[19] The order of arrangement of the molecules in the films of catalyst can alter the productivity of the reaction vastly, from almost no effect to a highly efficient reaction. A strong effect of temperature was also found, and the experimenters suggest that this was because very high temperatures disrupted the order of the molecules.

Polyunsaturated fats form a large part of the fat intake of Westerners. What the term refers to is the sequence of single (saturated) and double (unsaturated) carbon to carbon bonds in the fat molecules, somewhat like different kinds of beads on a string. A molecule having many double bonds is referred to as polyunsaturated. In addition, the spatial arrangement of the atoms attached to these double bonds can vary. A "cis" arrangement, which occurs in vegetable oils, makes for bulky molecules that are liquids at ordinary temperatures. If hydrogen is reacted with these liquid oils at high temperatures, some of the double bonds have their arrangement changed to the straighter "trans" form, so that the liquid oil becomes a soft or solid fat. Trans fatty acids are implicated in heart disease and possibly some degenerative diseases. Hydrogenation is very common both in manufacturing processes and as a food process, and polyunsaturates are used to make soft vegetable fats, often added to biscuits and margarine.

We all remember from our school days chemistry experiments that sometimes worked and sometimes did not. If they didn't work, it was usually the experiment that was blamed, not the theory. In fact, of course, every experiment "works," in the sense that it produces a result and therefore requires an explanation. Sometimes the failure of the experiment is blamed on not shaking the bottle properly, or on impurities in the flask. These kinds of explanations are really about sequence and order, but they are not strictly rational according to the known facts of chemistry, and they have the same place in the chemist's technology as "fudge-factors" (methods that, even though irrational, get things to work) do in electronics or mechanical engineering.

Jacques Benveniste, formerly of INSERM (Institut National de la Santé et de la Recherche Médicale), carried out a series of studies of the effectiveness of homeopathic medicines. He found that they retained their potency even at very high dilutions in water. The solutions were so dilute that there was almost certainly none of the drug present. He hypothesized that the drug could be effective at such dilutions only because the shape of its molecules had in some way become imprinted in the water. This so-called ghostly imprint hypothesis generated a huge outcry among those who thought such an idea superstitious.[20] However, it is known that sequences in fluid mixtures are difficult to disrupt once they are established; one such "lump" is known to physicists as a soliton or isolated wave. Is it unreasonable to suppose that the shape of the drug's molecules could in some way become imprinted in the pattern of the water molecules? Considerations like these take us into areas that are often dismissed as "fringe" science, but it is clear that we do not have as complete an understanding of the world as we sometimes suppose.

THE EARLY WORLDVIEW

Iteration and sequence, producing order, enter very deeply into the nature of the world. At one time this must have been much more obvious than it is now. Before civilization, humans would have been much more aware of the iterative processes surrounding them. Growth was more apparent, as was decay. Birth and death and the seasons of fertility and harvest brought home to people the iterative nature of things. Boredom must have been common as a result of the repetitive nature of daily life and its routines, which must have been without remission except for story telling and singing. In addition to being recreational, these activities would have served a valuable educational and cultural function. History and oral tradition in a time before writing and reading clearly necessitated the iterative process. Oral repetition was necessary for learning, and indeed went past this stage into what is now called overlearning. Knowledge was passed on by means of song and chant: some of these songs still survive, such as the old English folk song "Green Grow the Rushes O!" They are distinguished by iteration in number and repetition of the numbers on a cumulative basis. The form of the verses is 1; 2,1; 3,2,1; 4,3,2,1, and so on, so that 1 (the most important fact) is repeated twelve times, more than any other. Many people can still sing this song.

In a world where writing was either unknown or a rarity and information had to be remembered and passed on by other means, knowledge needed to be arranged in a way that would make it remain in the memory. One effective device was to build it into a poetry or prose structure naturally grasped by the human mind, so that the facts could be remembered by the individual and passed from one generation to the next.

THE ANCIENT WORLDVIEW

The early religions of mankind were centered around explanation, and primary in this explanatory scheme was their account of creation. In Egyptian theology the world emerged into order from chaos—Nun (or Nu), a name more concretely identified later with a goddess. Nun was a dark, watery deep containing a creative power. This hidden power was supposed at the first stage of creation to have been unconscious—the state of Nun did not know its own potentiality. Nun lay therefore somewhere between order and randomness in the scale of things.

To the ancient Egyptian mind it seemed clear that matter had always existed; in their view the idea of creation from nothing, even by a god, was illogical. Creation was thus the emergence of the properties of order from chaos. In other places, such as those known later by the Greek names of Memphis and Heliopolis, different versions of the creation story prevailed. In Heliopolis it was the sun god Re, and in Memphis the god of the earth Ptar, who created the first island in these turbulent waters. In any case, it was only at this creative moment that self-consciousness emerged, even in the god itself.

In this clear distinction between the order of the cosmos and the preexisting chaos lie the roots of the major world religions: Hinduism, Buddhism, Christianity, and Islam. In all of them a powerful sense of world order is apparent, often attributable to a single god, but again often existing simply as a background to the gods, one limiting their power and to which they were themselves subject. To the ancient Greeks the gods functioned against a scenery woven by the three Fates: Clotho the Spinner, Lachesis the Allotter, and Atropos the Inflexible or cutter-off of the thread. To the Romans these became the goddesses Nona, Decuma, and Morta. (Nun and Nona the fabric weaver probably become the Norns in Nordic mythology.) Even in Christian theology, where it seems almost blasphemous to question God's powers, theologians have maintained that He is indeed limited. According to St. Thomas Aquinas, one of the greatest Church Fathers, there are certain things that God cannot do; these include undoing the past, making a contradiction true, and willing His own destruction. These limitations on power imply a framework of law against which the Christian god operates. This is strikingly illustrated in Milton's *Paradise Lost*:

> Since by fate the strength of Gods
> And this empyreal substance cannot fail

the implication being that God, supposedly the creator of heaven and earth, is powerful solely because of His place in the fated scheme of things.

THE MEDIEVAL VIEW

By medieval times an inherited structure was in place, one that was not arbitrary but was, in the West, centered on a worldview based on Christian religion and the Bible. It is the arrangement of knowledge in this systematic way, amounting to a universal structure, that perhaps best distinguishes medieval from modern thought. Since the rise of rationalism, facts to be learned are no longer embedded in a universal structure but are grouped in subject areas in an encyclopedic way. To know a fact nowadays is to know something from one of many disconnected areas; in medieval times it was to know one aspect of a whole.

A book that influenced thought and scholarship for more than a thousand years was Boethius' *De Institutione Arithmetica* (On the Foundations of Arithmetic),[21] written in Latin after the fall of the Roman Empire. Boethius was from a patrician Roman family and in the sixth century tutored the new barbarian kings in the skills that had supported the power of Rome. One of these skills was the use of numbers, and Boethius set down the principles of arithmetic using numbers and their related geometry in a very modern style of mathematical thought and without the use of modern Arabic numerals. This in itself is a remarkable achievement. Roman numerals are not convenient for arithmetic operations: it is not easy to divide XXVIII by IV as a sum done on paper. However, Boethius showed how it is possible to work within such a system by understanding the properties of the numbers themselves as shown by their geometry.[22] He thus helped to lay the foundations of modern mathematics and also anticipated the relationship between geometry and arithmetic, rediscovered in the seventeenth century by Descartes and later elaborated by those working on the foundation of mathematics at around the end of the nineteenth century.

The ability to do arithmetic had far-reaching effects. The principles derived from arithmetic were used to structure knowledge in written texts and poems to make it more easily remembered. Texts written in this way contain a very high degree of arithmetic order, as has been demonstrated by David Howell. In medieval times Latin was the universal language of scholarship, and such highly structured works were composed by many Latin scholars of the period, who rivaled one another in the skill with which they used arithmetic structure to set down important learning "so that it might not fall from the memory." Using such devices, the seventh-century Irish scholar Mo Cuoróc maccu Net Sémon came to be known as "doctor of the whole world" in Rome. Another seventh-century Irish scholar, Aldhelm, wrote a letter in verse to a prospective student, trying to entice him[23] from Ireland to Canterbury. Howell shows how sections of this letter form a series of structures nested one inside another, so that the positions of words related by meaning are mirrored, and within each section further nested mirror images occur.[24] In addition, significantly related words fall at "golden points"—points determined by the golden mean or golden ratio.[25] The same ratio was used in the visual arts, for example, in the positioning of significant objects in a painting. It also occurs in the topography of living things, including plants and the human body.

Boethius also wrote a book on the art of music: music in medieval times was thought of as number in motion, the harmonies used and the changes between them corresponding to number theory. The entire worldview of the times was geared to the human understanding of a divine plan embracing every phenomenon of nature.

THE EARLY SCIENTIFIC WORLDVIEW

When science began to emerge as a systematic method it was seen at first as an extension of and an accompaniment to the traditional methods of gaining knowledge about and control of nature, not as a replacement. Two of the people who were to be key figures in the emerging scientific worldview, Isaac Newton and Francis Bacon, did in fact see science as an extension of existing ways of knowing and carried on their scientific activity alongside their religious, mystical, alchemical, and magical practices. Figures such as the mathematician and astrologer John Dee, now often considered charlatans, were in their time as respectable as scientists as was Isaac Newton, also a mathematician, astrologer, alchemist, and theologian.

The Age of Reason changed this outlook completely. Acceptable canons of knowledge were laid down, even if this meant classifying some phenomena as apocryphal—specifically those phenomena considered miraculous. Along with the rejection of miracles[26] went rejection of the religious world order that they were seen to be supporting. The Encyclopedists of the French Revolution put the new rationality into effect. The idea of an encyclopedia is profoundly atheistic, suggesting as it does an arbitrary, manmade arrangement of knowledge.

THE MODERN SCIENTIFIC VIEW

By contrast, the post-nineteenth-century worldview is that random natural processes have shaped the world and living things, including humanity and all its actions. This belief has underlain almost every area of modern science, especially biological science, where it is most entrenched. We will look in greater depth at the consequences of this view—what I call the random-selection worldview—in later chapters.

This change in scientific thinking was the product of the nineteenth century enlightenment, which originated with Charles Darwin. A religious man, at least outwardly, Darwin suppressed for many years his intuitions about evolution because he thought, quite rightly, that they would prove fatal to the very beliefs he professed. His champion, Thomas Huxley, inaugurated a new view of science, one based on the blind operation of natural forces, the implications of which could only hasten the doom of an already crumbling worldview. Following soon after the Darwinian idea of the random selection processes of nature came other new discoveries: relativity, the uncertainty principle, and the unconscious. The total effect was to render man all but helpless in a world where he was no longer a willing actor, but rather a creature of forces he could barely control. This was the bleak scientific faith that dominated twentieth-century thinking and made in many ways a fitting accompaniment to the meaninglessness of that century's mass destruction.

Insofar as science has become the guiding principle of our times, and insofar as the idea of randomness has directed science, I believe that progress for humankind is obstructed by the paradigm of random selection. How can the situation be retrieved? Heraclitus of Ephesus frequently mentioned the idea of the *logos*. The logos is what is common to all. He says

> *Those who speak with sense must rely on what is common to all, as a city must rely on its law, and with much greater reliance: for all the laws of men are nourished by one law, the divine law; for it has as much power as it wishes and is sufficient for all and is still left over.* (FRAGMENT 114)

If we are presented with some phenomenon in biology or in society, we must ask ourselves what these things have in common that gives them a common function. Iteration is clearly a part of the logos, because Heraclitus says

> *Beginning and end in a circle are common.* (FRAGMENT 103)

Despite appearances, there is unity in the world; even, perhaps especially, in opposites:

> *They do not apprehend how being at variance it agrees with itself: there is a connection working in both directions, as in the bow and the lyre.* (FRAGMENT 51)

In the end, the logos or system rules determine the system goals and hence the system values. Without a grasp of the logos, the solution will not be possible, but even with it, it will not be easy, for

> *The real constitution of things is accustomed to hide itself.* (FRAGMENT 123)

In the rest of the book we will see how sparse in some areas is the understanding that twentieth-century science has left us. Information and order are powerful concepts; however, they do not enter often enough into the thinking of life scientists. Chance puzzles people, while a sense of coincidence fills their lives and leaves them with a sense of unexplained mystery. Biology can describe a great deal, can manipulate even more, but still understands very little. Mathematics is considered by many to be on the firmest of all foundations and unquestionable in its truth; however, as we shall see in chapter 10, this is far from the case. In all these areas the role of sequence is vital, and in all of them sequence is produced by iterative processes.

As we explore these processes we shall find that the traditional scientific model of the world has become inadequate in many important respects and that we are forced to abandon it and look at things in a completely new light.

2

THE CRISIS IN BIOLOGY

This order did none of gods or men make, but it always was and is and shall be: an ever living fire, kindling in measures and going out in measures. (FRAGMENT 30)

BIOLOGY IN ECLIPSE

It may seem odd to say that biology is facing a crisis. At a time when the successes of molecular biology are proclaimed in every quarter, surely biology has never prospered more? What criticism, it might reasonably be asked, can be made of the science that is apparently the most successful, that is the basis of the boom industry of genetic engineering, that has led to the understanding of the cause of many genetic illnesses, and that is the great hope for the future of medicine? The answer is: none. However, biology has done few, if any, of these things and has contributed almost nothing to their success. The truth is that the science of biology has never been weaker; it is in crisis, and that crisis is made worse, not better, by the extraordinary success scientists have had in discovering the structure and basic mechanisms of the genetic material.

The elucidation of the structure of DNA and the structure and function of the other nucleic acids and intracellular components has certainly been an immense triumph, but not for biology. These achievements have been brought about, not by biologists, but by chemists. Biology and biological knowledge are thought irrelevant to these subjects, so much so that a good grade in chemistry is now the primary qualification needed to enter a degree program in molecular biology or genetics. Biology is a subject that (at least in Britain) is not required, and this alone says almost everything about the crisis in modern biology. Insofar as genetic science explains living things, biology has no part in the explanation. To put it bluntly, biology has been expelled from its own territory.

Instead, the ground is occupied by scientists from an alien culture, who need to know nothing about biology in order to do their work.

The result of this incursion is the lack of any biological influence on the thinking going on in genetics, the science that must be at the heart of morphology, development, and evolution; vital issues, surely, for mankind's understanding of itself and of its predicament.

How did this happen? Surely some strange and unique process must have been at work to bring about the expulsion of the science of life from its own workplace? Not so; in fact, the problems faced by biology today are the same as those faced by other sciences, but they are more acute paradoxes because biology deals with the most complex of all systems—those that we class as living things. What is to blame, once again, is the failure of the prevailing approach—the random-selection paradigm—and its constituent elements.

Biology has at the moment no clear direction, it is making little progress toward its proper goal, and it is merely being distracted from that goal by inappropriate theorizing. To justify these statements I shall start by considering the relationship of biology to other sciences with which it is interconnected. Currently, biology is hamstrung by two approaches that, as they stand, will not advance it at all: molecular biology and genetics, on the one hand, and neo-Darwinian theory, on the other.

THE SUCCESS OF MOLECULAR BIOLOGY

As everyone knows, James Watson, an American biologist, and Francis Crick, a British physicist, assisted by less well-known but equally important others, in 1953 unraveled the double-helix structure of the chemical DNA. This single discovery brought the promise of order and progress to a notoriously vague and erratic field. Until then, attempts to unravel the secrets of biology using the techniques of "hard" science had failed signally to work. The three cardinal doctrines of causality, atomism, and randomness, which had served the rest of science so well for hundreds of years, had hardly helped biology at all. Biology was considered such a disorderly and ill-defined subject that it was not thought to be at the stage when a truly hard approach would pay off, though the day was eagerly awaited when exact scientific methods would find their application in what was considered to be a very soft area.

All this suddenly changed when the structure of DNA became known. The success of scientists in unraveling the makeup of the long, complex molecules that are the basis of life led to a discipline dominated by research techniques and thinking appropriate to hard science. It used to be said that such an approach wouldn't work: that living things do not lend themselves to an analysis in terms merely of cause-and-effect events. But this criticism was now seen as futile, because the approach clearly was working.

Or apparently working: the trouble is that the type of theorizing and the kinds of research technique it leads to are those of chemistry and not biology. Since the dominance of chemical rather than biological thinking has produced such rich rewards, it is supposed that it will solve all problems. This attempt to fit biology into an inappropriate framework has led to only the molecular biology approach being considered acceptable while all others are neglected.

CHEMICAL THINKING

By chemical thinking I mean thinking that is based on the view that chemical processes are the causal agents in biological organisms. This model takes the view that once you have identified the components of the system and their possible interactions, all the problems have been potentially solved. The picture of life built up in this way is roughly as follows: the molecules RNA and DNA, which are found in the cells of all advanced living things, are the templates according to which the organism is built. Groups of bases on these long molecules are translated into amino acids, and these amino acids form polypeptide chains, which either alone or together with others, constitute protein molecules. Living organisms are those arrangements of proteins that are allowable under natural selection: if a given arrangement is successful, it survives; if it is not, it perishes.

This model, although in principle aimed at explaining complexity, in fact often leads the scientist to reduce everything to simple cause-and-effect chains of chemical events. This in turn leads us to suppose that, although these chains of events are rather complicated ones to discover and to understand, when this has been done, all the outstanding problems will be solved. The model holds out the possibility of understanding genetic disease and curing it; it offers the promise of constructing new types of organisms engineered to have certain desirable characteristics; and, hopefully, it will help supply the missing link between natural selection and what we see in the world of living things. According to this view, hard science has (potentially at least) once again conquered the central ground in a new area—the previously disorderly one of biology.

Notice, however, how much the model leaves out: the questions it cannot, or does not, attempt to answer. How does a cell function to preserve itself? What are its necessary interactions with other cells? How did the nucleic acids arise in the first place and what is the function (if any) of the noncoding (silent or redundant) sequences within them? Why do they contain numerous and sometimes extended repeat sequences? Why does natural selection not enable us to predict, or even account for, the form of an organism? Why does evolution seem to proceed toward ever greater levels of complexity?

These questions can be answered only at a *biological* level, and chemistry by itself cannot advance our understanding of them one jot. Chemical laws, which apply to the combination of substances, cannot tell us anything about the purposes of animal behavior, the organization of a group of cells, or even the working of one living cell. It is as if we were to try to understand a television set by means of electronics, leaving out its functioning, the meaning of the messages it carries, or the social context of its use; or as if we were to try to understand human beings solely by means of biology, ignoring the study of psychology or of social groups. We cannot gain any understanding of the fundamental features of living things solely at a chemical level.

EMERGENCE

To make an analogy, a similar situation exists in psychological research. Psychology is a very broad science: at one end of it we find social psychologists studying the behavior of people and at the other end we have neuropsychologists studying the behavior of neurons. Although these two approaches are very different, they are both recognizably tackling psychological questions, albeit in different ways. However, there is always a hope in the minds of the neuropsychologists that one day, because theirs is the most fundamental level of investigation, all behavior will be explicable in terms of neurons and their interconnections. Is this so? Can all the problems of psychology ever be solved at the neuronal/neural level? The answer is, no more than the problems of biology can be solved at the molecular level, because without an understanding of what is going on at the behavioral level, the predictions of neural models are meaningless. As Coles[1] made clear in his presidential address to the Society for Psychophysiological Research in 1989, only a marriage of psychophysiology and psychology will yield an understanding of how the functioning of the brain explains actions and experiences. And yet there are those, like Michael Gazzaniga, who would like to see the teaching of Freudian psychoanalysis prohibited and psychology departments closed down, as a waste of time.[2]

The key distinction I want to make is one made by Elliott Middleton, who has suggested that the boundary between one science and another be drawn at the highest level that can be described within its vocabulary. Middleton proposes that the unit of analysis in any given science should be what he terms the "emergon." An emergon is the most complex object that can be constructed in the language of the science next to, but lower in complexity than, the one in question. The emergon is, if you like, the "atom"[3] of the next science up. For example, the elements (like oxygen or iron) are the most complex objects capable of description within atomic physics and so are the basic constituent units of chemical science: the element as a unit is thus the "atom" of chemistry. The properties of these emergons cannot be fully described within the language of the science from which they emerge. For example, the properties of elements or the results of their interactions cannot be arrived at within physics; they emerge only as a result of chemical observations. The hardness of iron or the wetness of water are not describable by the language of physics; to describe them one needs to move to the next and higher level of science—chemistry.

According to such a nested picture of the sciences, the inadequacy of chemistry as a description of biological organisms is clear. Chemistry just does not contain the constructs or concepts required to describe the phenomena of biology. Of course it could acquire such a vocabulary, but to that extent it would then no longer be simply chemistry; it would be a combination of two separate sciences—chemistry with biology grafted on to it. To the extent that chemistry alone is being used in our thinking, it is incapable of explaining the phenomena of life.

KINDS OF EXPLANATION

The second reason for the inadequacy of chemical explanations in biology is the *type* of description offered by chemistry. Such descriptions, although they are long and complex, are simply chains of events. To give such an account amounts to little more than conveying the primitive notion of one thing "causing" another to happen. Molecular biology has notably failed to provide a statement of functional relationships that might begin to explain the way an organism works. Its explanations are couched in terms of causes and effects, typical of a primitive stage of scientific understanding. Explanations of this type lack the precision offered by a science in which the description is in terms of functional relationships between continuous variables.[4]

It would be quite wrong to imply that mathematics is not used in molecular biology—there are mathematical models of most aspects of molecular interactions. But molecular biology cannot by its nature produce any account in terms of math or information theory of the functioning of living things, when information is the most fundamental property of an organism. There is a dividing line between a science that has been mathematicized and one that has not (or not yet), and the hard science of chemistry is in fact not hard enough to enable biologists to cross the line.

There are certain modes of discussion common in the analysis of gene function that illustrate this point. Processes that occur naturally and can also be experimentally induced are methylation, acetylation, and phosphorylation. Gene transcription may be effectively switched off by the addition of a methyl group in the appropriate nucleotide position, while adding an acetyl group or a phosphate group to the histones can facilitate gene expression.[5] In discussion, this can lead to a purely causal picture of the kind referred to above. For example, in an account[6] of the formation of nodules on the root meristem in legumes, the discussion is in terms of the genes "regulating cell fate." The gene HAR1/NARK, which appears to have the same function in different legumes (HAR1 in *Lotus japonicus* and NARK in soybean), is called "a regulatory gene that normally limits nodule numbers." The impression given is that the gene has somehow accounted for the feature (nodule number in this case), but exactly how is left unclear.

"Signaling" is also used often as a metaphor in descriptions of gene function. A signal may be anything from a simple on/off switch to a stream of information. All too often it signifies a simple switching process, yet the ambiguity in the use of the metaphor again hides this fact.

The unraveling of the human genome is a triumph of molecular biology, but to give the impression that all is now explained, or even potentially explained, is quite misleading. What is lacking from a bare description of the genome is any approach that might explain the significance and function of its structure.[7] To say that molecular biologists are still trying to understand life at the level of chemistry (or even physics) is simply another way of saying that there does not yet exist a truly biological level of science, that there is nothing scientifically available at the level of the living organism for them to use. Biology, in the sense of molecular biology, is still fundamentally descriptive and there is as yet no way of fitting the descriptions together to form a holistic and meaningful picture.

THE NEO-DARWINIAN DEBATE

At the other end of the biological spectrum of complexity from the genomic is the evolutionary. Apart from its fascination as a subject, which arises largely from the fact that the story of evolution is also the story of ourselves, evolution is studied for many reasons. If we could understand how living things became the way they are, we might understand many other things about them, such as how they got their present forms and functions, what is likely to happen to them in the future, whether they always tend to become more complex, and how they are interrelated. All these questions demand a *biological* approach, an understanding of the laws governing the evolution and development of living things.

This biological approach to evolution has long been impeded by a particular case of the random-selection paradigm—Darwin's theory of evolution by natural selection—which forms the "official" basis of biological thinking on this subject. The random-selection paradigm as applied to evolution is so unsatisfactory that it would be hard to find an evolutionary biologist who could say, hand on heart, that they were happy with it. Yet in spite of all the attacks made on it, neo-Darwinism survives, like the creatures that it explains, because it is the fittest of the available theories. This is actually to understate the degree to which neo-Darwinism is accepted by the scientific community. It is the *only* available theory; it has no competitor.

Neo-Darwinism is the theory that evolution is brought about by natural selection based on random genetic mutations. For natural selection to occur, there must be a number of conditions: there must be variation (differences in traits or attributes) between members of a species; the traits must be related to the likelihood of reproduction (so that possessing the trait increases or decreases the probability of reproducing); and the traits must be heritable (able to be passed on to the offspring). The source of the variation is assumed to be random mutations in the genetic material. If these conditions hold, then it is maintained that there may be predictable changes in the distribution of the traits over time (evolution). It is the essence of neo-Darwinism that small, undirected changes that make for reproductive success will be passed on and will accumulate to produce new species and that natural selection (and natural selection alone) is the force that makes for evolutionary change.

First we must make a very basic distinction between belief in evolution and belief in neo-Darwinism. People who question aspects of neo-Darwinism are often labeled as denying the theory of evolution, but this is absurd. The neo-Darwinian theory is but one possible explanation of the way in which evolution could have taken place. Fundamentalists deny that humans have evolved; but to believe that evolution has occurred it is not necessary to be a Darwinian.

SELECTION VERSUS PRODUCTION

Next we ought to distinguish between accepting the importance of natural selection and accepting it as the basis (let alone the sole basis) of evolution. There can be little question of the importance of natural selection as a mechanism; what distinguishes the neo-Darwinian position is the doctrine that it is the only effective way in which new types of living thing are produced. That natural selection is the only mechanism of change—that evolution is driven solely by natural selection—is the distinctively neo-Darwinian position. Given that natural selection operates, as everyone with knowledge of the facts agrees, is it reasonable to assume that it explains evolutionary change?

The arguments against Darwinism are many; some of them are very old, and were either made by Darwin's critics at the time, or even suggested by Darwin himself in considering what might be said against him. The fact that some of these arguments are old does not mean that they have been answered: many have simply been ignored or shuffled aside.

The philosopher Karl Popper proposed that science does not proceed by verification (proving that a theory is valid) but by refutation (proving that other theories are invalid). While this may be a useful proposal about how science *ought* to work, it is not how it does work. What happens is that, if a theory is successful in explaining a lot of facts, then any discrepant facts are either said to be irrelevant or simply ignored. It is only in its later years, when the theory is failing in many obvious ways, that the awkward discrepancies may be reconsidered and the old theory may finally be dismissed and replaced by a newer one that explains them.

Scientists are, in this, just like other human beings: they form broad concepts that fit together into a cognitive scheme. It does not have to be a consistent scheme, and usually it isn't. In the philosophy of science these concepts are called *paradigms*: a paradigm is so broad that it will meet the need for explanations in a wide area of observations. A successful paradigm may leave out a lot of phenomena (or even conflict with observation), but that will not necessarily lead to its downfall. Only when a new paradigm arises to overthrow the old is the old finally seen to be inadequate.

"ONE LONG ARGUMENT"

Over the last few years many cracks have appeared in the orthodox paradigm of how evolution happened, which for want of a better term I shall call neo-Darwinism. Neo-Darwinism is a synthesis of three areas: it combines Darwin's theory of evolution by natural selection, Mendel's laws of inheritance, and modern findings from molecular biology. The neo-Darwinian view is that small, undirected, essentially "random" changes that make for reproductive success are passed on and accumulate to produce a new species. In this way the present forms of life have (the theory says) evolved over a period of some two and a half billion years, while the eukaryotic life forms have evolved in less than five hundred million years and humans have evolved from their precursor in less than a million. According to the existing theory, these changes have occurred simply as a result of a series of random mutations having led to an enhanced probability of reproduction of viable descendants carrying the mutation. What criticisms have been made of this point of view and do they amount to a rebuttal?

The main argument against Darwin's view was, and remains, the absence of those specimens of the supposed intermediate species that, according to the theory, should at some time have existed. Under the Darwinian position, during the course of evolution a series of small changes is supposed to have taken place, resulting eventually in complete separation of a new species from the preceding one. Therefore, between any two species of the same genus, there must at one time have been a third, intermediate, species. Often, however, no remains of such a species have been found. The classic example is the "missing link" between humans and the other hominids such as Neanderthals from which they might have evolved. But almost any line of differentiation between two species will give rise to the same problem, the dividing line between dogs and wolves, say, or between two different kinds of mollusk. It seems that differentiation between species often occurs without any apparent intervening specimens.

The explanation usually given for these missing pieces of evidence, and the one that was given by Darwin himself, is that there is a very low probability of finding the fossil remains of an intermediate species. The intermediates, it is argued, were less viable than either those from which they sprang or those they eventually produced. Because of this, the intermediates quickly evolved further and consequently left behind fewer fossils than did the end-point species. Those few that might have been left have not been discovered, partly because of their rarity and partly because of geological changes that have obliterated them. Geological evidence can be produced both for and against this suggestion. However, the absence of intermediates certainly does nothing to support the neo-Darwinian position.

On the other hand, there are specimens of evolved species that are known, or appear, to have changed radically with no intermediates. For example, the homeotic (structural) mutations of the fruit fly *Drosophila* are sudden and without gradation. In the coelacanth *Latimeria* the median fins appear to have changed shape suddenly and radically early in the history of the group.[8] In his book *Sudden Origins* Jeffrey Schwartz surveys such sharp changes, where a new species seems to appear literally overnight.[9] These data suggest a different model—one based on sudden, mutation-driven changes in morphology rather than gradual change and, as we shall see, other evidence favors this position as well.

HOW LONG IS A PIECE OF DNA?

The second argument is that the time scales involved in evolution, although very long, do not appear to have been long enough for the process of gradually accumulated change to have produced creatures of the complexity that we observe today. The whole of evolution from its beginnings to the present day took about three billion years. In this admittedly very long period, many changes would have had to occur and be selected to account for the evolution of advanced forms of life from a posited prokaryotic organism.

How many changes were involved? It is almost impossible to estimate a numerical value for the rate of change or the likelihood of its occurring, because there is currently no way of quantifying morphological features nor of estimating numerical values for the other variables. One possible approach, notoriously taken by Fred Hoyle, is to ask how long it would have taken to produce a given protein from its constituent amino acids—the building blocks from which they are made.

There are 20 of these amino acids in proteins.[10] The total time taken to construct a protein depends on its size and what is assumed to be the unit of time needed to place an amino acid in its appropriate position in the sequence. Consider a protein requiring about 1,000 amino acids and suppose that it takes a second to place one of them. The correct one out of the 20 has to be selected and placed in a given position in the sequence, so if it took one unit of time to place one amino acid, it would take 20 times as long to place 2, 400 times as long to place 3, and 20 to the power 1,000 times as long to place all of them. At this rate it would take over a thousand billion years to create this one protein, unless some other principle were at work. And this is only one of the many thousands of proteins required to constitute an organism. The proteins themselves would have to be capable of working together in the organism, and the resulting organisms would have to reproduce in a way permitting natural selection to operate. Unless some organizational principle is applied, changes would have to occur and be passed on by reproductive success incredibly often over a sustained period to arrive at the results we see today.

A frequent response of the neo-Darwinian to this kind of argument is that the interaction between different genotypical features to produce the phenotype—so-called epigenetic interactions—are not taken into account. This is surely the case; but it is exactly the nature and effect of these epigenetic interactions that is in question and remains to be determined.

MONKEYING WITH EVOLUTION

It has been said[11] that a horde of monkeys typing on keyboards would eventually produce the works of Shakespeare. Put like this, evolution by natural selection looks not only natural but inevitable. But a little calculation soon shows that the time the monkeys would require exceeds the age of the universe by a factor of many orders of magnitude. So the monkeys (which are a surrogate randomizing mechanism) are of little help when it comes to supporting the idea of evolution by random variation alone. Richard Dawkins, in his book *The Blind Watchmaker,*[12] suggests a solution to this problem by likening evolution to a type of computer program called a genetic algorithm. Genetic algorithms are programs that can evolve toward the solution of a problem by taking two attempts at the solution and combining (or breeding) them to get further, and hopefully better, solutions. Genetic algorithms are vastly superior to brute force algorithms at achieving their goal, and Dawkins shows how one such program, which he wrote himself, can produce the sentence "Methinks it is backed like a whale" from Hamlet. This sentence is composed of 34 letters, including spaces. If left to print out all the possible combinations of 34 letters without any directing algorithm, a computer (or a set of monkeys) would almost certainly never hit upon this sentence, let alone the whole play. But Dawkins' genetic algorithm can do it in less than fifty steps.

Dawkins uses this as an argument for the power of evolution by natural selection acting on random mutations. But the crucial difference is that a genetic algorithm is used by someone (in this case Dawkins) to work toward a known goal. At each stage, the string of letters produced by one of the versions of the program is compared to the target sentence, and it is on this basis that the candidates for further breeding are selected. The implication of Dawkins' argument, if taken at face value, is that life knows, and has known all along, where it is going. Such a goal-directed or teleological progress of evolution is a highly unorthodox assumption in neo-Darwinian theory.

A further difficulty, related to the same point, is that it is hard to see how gradual changes could have been continuously advantageous to the organism that carried them. The reproductive value of precursor features that must have existed at some stage, such as partially developed organs, is by no means obvious. A classic objection made to Darwinism since its early days was that it would be very unlikely that mammalian sight could have evolved in this way. It was often phrased as the rather absurd question "What use is half an eye?" Some support for the Darwinian position had been found because a computer simulation[13] has demonstrated how an eye could evolve in only a few hundred thousand years. Dawkins, in his later book, *Climbing Mount Improbable*,[14] made much of this study, but even as little as a hundred thousand years still seems a very long time for the uninterrupted selection of mutations toward a single goal.[15]

MIMICRY AMONG INSECTS

An interesting case of evolutionary change is so-called mimicry among insects, particularly butterflies. Unrelated species of butterflies sometimes share common patterns or parts of patterns on their wings, or they share other characteristics such as mode of flight or sitting posture. Often when this situation occurs, one species is more palatable than another to their predators, or one species is palatable while the other is actually distasteful. Those species of butterflies that are unpalatable are avoided by predators (birds), who either are instinctively programmed to avoid them or have learned to do so.

In 1861 William Bateson put forward a theory concerning these butterfly species that resemble one another. Bateson proposed that the similarity in appearance had arisen because it was advantageous to the more edible butterfly to appear like the less edible (called the "model" species) and that the process of natural selection had ensured the survival in greater numbers than otherwise of variants that resembled the model. The palatable species benefit greatly from this resemblance because they get a "free ride," as it were, in the survival stakes. The question is, how did the resemblance (referred to as Batesian mimicry) arise? The detailed arguments that ensued about this situation cast doubts on the doctrine of gradualism.

Batesian mimicry is at first sight a double-edged sword. If two species A (palatable) and B (unpalatable) resemble one another sufficiently, then, although A gains by the resemblance (by looking like a member of a less-edible species), B loses by it (because it looks edible to birds that have been eating the palatable species A), even though it isn't palatable. This would lead to Bs being attacked before the birds learn that their appearance belies their taste and that they are not As at all. So it seems that any gradual process that could make A move toward B in appearance would act equally to move B away from A.

Much, of course, depends on the numbers of butterflies in the two species. If there are more butterflies in species B than in species A, then an individual member of B loses less by any resemblance than an individual in A gains. It was agreed that Batesian mimicry worked best when the palatable butterflies were relatively few in number compared to the unpalatable.

However, there are also examples of species, *both* numerous and *both* unpalatable, that also resemble one another. This led Müller in 1879 to propose a different explanation of mimicry: that the benefit in the case of two unpalatable species was mutual, so that both species gained by sharing a common warning sign. This involved the birds learning, slowly and possibly with mistakes, that it did not pay to attack either species. Under these special circumstances of selection, the two species could, Müller suggested, converge in appearance in a mutually beneficial and even cooperative way.

But similar arguments emerged in Müllerian mimicry as in Batesian mimicry. If two species A and B initially looked very different, and mutants of A looked more like Bs than before, it was argued that the A mutants had lost the advantage of looking like As before they had gained the advantage of looking like Bs, and so would be *more* at risk of predation, not less. Only if the mutation was a sudden jump could As gain an advantage from it.

G. A. K. Marshall in a paper[16] written in 1908 considered this problem. Marshall argued that if there are two species, equally unpalatable, the less numerous one would, if anything, evolve to mimic the more numerous, never the other way around. This occurs because, if the species initially looked different, the birds would eat proportionally more of the less numerous species before learning that it did not pay. Any of the more numerous species A who through mutation came to resemble the less numerous B would be putting themselves at unnecessary risk by pretending to look like Bs. "When the unpalatability is equal" went Marshall's conclusion, "the less numerous species will be attracted by the greater, but the greater will not be attracted by the less." In Müllerian mimicry, as in Batesian, an approach would only take place if the species were very different in numbers.

In this situation, where both species were unpalatable, there would be a benefit for both species in appearing like the other. The crucial question is, how did the resemblance come about? If it was by a gradual process, then each step must have been advantageous. But Marshall argued that there is no apparent advantage in looking like *neither* of these nonpalatable species; there is only an advantage in looking like one or the other. It follows that there is no continuous selective pressure that would take B toward A, much less vice versa. So the transformation must have taken place in one leap.

"SUDDEN MIMICRY"?

R. A. Fisher, the geneticist and statistician, in his book *The Genetical Theory of Natural Selection,*[17] challenged this conclusion and argued in favor of gradualism. One of his grounds was that the criterion for evolution was fitness, in the sense of genetic survival, and that any variation that conferred greater fitness on either the individual *or its close relatives* would persist. The individual that was attacked and mistakenly either eaten or left for dead might not pass on any benefit from its coloration. However, because its relatives resembled it genetically, the species as a whole did benefit from the ruse. (This argument, from an extended concept of fitness called inclusive fitness, smacks of the one used later by gene evolutionists like Trivers and made famous by Richard Dawkins in *The Selfish Gene.*[18])

Arguments, entirely theoretical, raged. Some argued that a mutating species lost the advantage of looking like itself before it had acquired the advantage of looking like the model. Others maintained the exact opposite: that it retained the advantage of looking like itself and at the same time acquired that of looking like its model! This sort of debate, so appealing in Victorian times, can hardly resolve a problem as complex as mimicry. There are too many variables: the appearance of both species, the supposed mutation rates, the behavior of the predators, the possible role of instinct as compared to learning, the effect of poor predator memory, and so on.

So little is actually known about mimicry that cases sometimes appear to be of the Batesian type but on closer examination have been found to be Müllerian! Ritland[19] tested the assumption that the Florida viceroy butterfly (*Limenitis archippus floridensis*) was a palatable Batesian mimic of the purportedly distasteful Florida queen (*Danaus gilippus berenice*). Surprisingly, he found that the Florida viceroy was even more unpalatable to birds than the Florida queen, and that the supposed Batesian mimicry was Müllerian in nature.

Mallet, in a series of papers,[20,21] has discussed the paradoxical nature of Müllerian mimicry. The adaptations concerned involve changes in only a few genetic loci, and multiple differences frequently appear in the same species in different geographical locations. The psychology of the predator is crucial. On its learning everything depends, since this forms the basis of the presumed selection. Mallet suggests that true Müllerian mimicry may be much rarer than is supposed.

As with other models of complex systems, this is an area in which computer simulations can help. In addition to the changes in the prey species, the behavior of predators (the birds) is very important. How quickly do they learn, if they do learn, that insects are noxious? How well do they remember? Computer simulations by Speed[22] have recently shown that the memory of the bird is a crucial and extremely sensitive factor in modeling mimicry. As the author remarks, their simulations show just how sensitive relatively small changes in behavioral rules may be to the role of predators.

In 2001 Durrell D. Kapan[23] did an experimental test of Müllerian mimicry by releasing two kinds of butterflies, morphs of the same species (*Heliconius cydno alithea*), into the habitat of two other species that they resembled. One morph of *H. cydno* was yellow and one white: the white morph acted as the control in the habitat of the white-patterned species *H. sapho candidus* and the yellow morph acted as the control in the territory of the yellow *H. eleuchia eleusinus*. The survival patterns of the released specimens fitted a Müllerian model in which great benefit accrued to the mimicking control butterfly, in comparison to the disparate, experimental specimens. However, this experiment cannot show anything about change in the species themselves. And as Kapan points out, the very existence of the morphs is contrary to what Müllerian theory predicts! The potential adaptive benefit should have caused the species of *H. cydno* to evolve toward a single monomorphic type rather than two types. Mimicry is evidently a complex phenomenon, but it altogether fails to present that clear picture of Darwinian gradualism that it was supposed, for so long, to illustrate.

SPECIES DISTRIBUTION

A different kind of objection to Darwinian gradualism was raised by J. C. Willis in his books *Age and Area*[24] and *The Course of Evolution*,[25] based on the observed distribution of species over land areas. This distribution is very uneven, so that there are old, long-established species inhabiting very large geographical areas while more recent species are confined to smaller areas. Moreover, these areas overlap, so that the two species, the precursor and the descendant, coexist in the same environment. This casts doubt on the idea that it is selective pressure that has caused the emergence of the new species, or that selective pressure maintains it. Why should a new species have emerged if the old one is still adapted well-enough to survive in the same geographical area? Why should the old species have survived if selective pressure favors the new one?

Willis points out that there are many small species (i. e., species with few individuals) and only a few large ones (those with many members). This distribution is more consistent with mutation as a cause of variation than with natural selection, because it suggests that the smaller species have evolved as mutations from the larger ones. Indeed, as Willis says, it is really an inversion of the prediction of natural selection. Under a selectionist interpretation, the small species should be few in number because either they are unsuccessful adaptations (they are dying out) or they are beginning to emerge as new species, in which case it is unlikely that we would be here just at the point of their emergence and thus able to observe so many of them! They must therefore represent equally viable solutions to the problem of survival, but are more numerous because they have been produced by the mutation of the large species.

The scarcity of more numerous species points to the influence of a mechanism other than natural selection: it suggests that, in many if not most cases, it is *mutation,* not natural selection, that is the basis of change without any selective pressure necessarily being exercised. Mutations occur to individual members of a species. While most mutations are lethal, some are harmless and can very occasionally give rise to new species. The new species may or may not be as well adapted to the conditions as the originating species, but it is at least adapted well enough to survive. The two are therefore found side by side: neither is disappearing and the new species is not there as a result of selection.

The geographical distribution of different sizes of species is also inconsistent with natural selection as an agent of change. Many of the small species are to be found on mountains and also coexisting with the larger and more widespread species to which they are related and from which they are presumably descended. This is not in accordance with natural selection, as already pointed out, since, if the new species had been produced by this mechanism, the old species should have died out because it was not well adapted. But such a distribution is entirely consistent with mutation as a source of variation, since it is in mountainous conditions that mutations are most likely to occur, because of enhanced cosmic particle bombardment and more exposure to ultraviolet light.

Willis also points out that it is often very difficult to see the particular value of botanical variations: why should it make any difference how many stamens a flower has? Why should it be necessary for plants to have their leaves exactly opposite one another on the stem? As we shall see, questions like these lead us in a completely different direction. While mutations are admitted by neo-Darwinian theory to be the source of variation, it is maintained by orthodox theory that mutations are essentially random or undirected. However, there is not, as one might think, a lot of variation between species, but surprisingly little variation. It may look as though there is a lot of variation, but the apparent variety is due to variations on a common set of themes. D'Arcy Wentworth Thompson, in his classic book *On Growth and Form,*[26] shows how three different skull shapes—human, chimpanzee, and baboon—can be obtained from the same basic form by transforming the coordinates on which they are drawn. The same technique can be used to metamorphose related species of fish into one another, leaving their features, including fins, eyes, and other body features, appropriately relocated. Thompson gives many other examples, enough to convince all but the determined skeptic of the broad applicability of his method to natural form. These coordinate transformations are not arbitrary, but consistent. They suggest an underlying mathematical rule linking species; a rule that Thompson spent many years trying to elucidate.

LIFE AND REGULARITY

There is in fact one single, common form that is possessed by a very wide range of plants and animals: it may be best described as dendritic or tree-like.[27] The dendritic form is displayed, for example, in the branching of plants and in the forked limbs of animals. The internal systems of animals are also dendritic—the pulmonary system, the circulatory systems, and the nervous system—so that this one form characterizes the bulk of all advanced species.

An interesting feature of the almost universal dendritic form is its symmetry: the left and right sides of animals are approximately symmetrical and plants possess even more axes of symmetry. Furthermore, the upper and lower parts of plants and the fore and hind parts of animals also have a rough symmetry. This has the result that the amount of information required to represent an advanced organism is much less than might have been supposed from the number of cells in its body.[28] Such organization of information does not suggest something that has evolved in a random manner.

Many difficulties for the neo-Darwinian position arise from this phenomenon of the remarkable regularity of organisms. Take the example, which Willis cites, of leaves growing on opposite sides of a stem. This can occur in one of two ways: two leaves can be either exactly opposite (100 percent opposite) or evenly spaced (0 percent opposite). Since evolution has occurred, there must have been originally one form: let us assume it was 0 percent opposite, and the 100 percent opposite case has evolved from it. This must have happened, according to neo-Darwinism, because of selective pressure to change. Now if this were so, there should be some cases observed that are not quite 100 percent opposite but, say, only 95 percent or 98 percent. However, there are none to be seen. If the leaves were slowly "sliding" along the stem under some selective pressure, then, when the adaptation had become 97 percent, the selective pressure would have been so reduced that this arrangement should have been stable. Why has evolution done its work so thoroughly in this case when in so

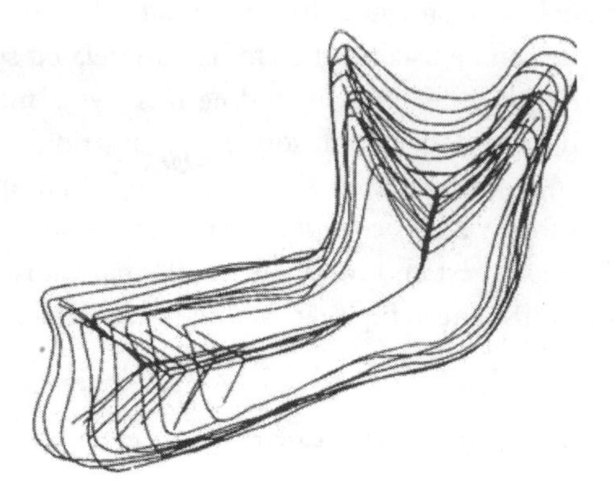

FIGURE 2.1 Invariance in the triple angle of the human mandible with age. Shown are Broadbent–Bolton standards for nine different ages overlaid at the point of maximum curvature. Each standard is averaged from several thousand children at ages 1, 3, 5, 7, 9, 11, 13, 15, and 17 years. The triple angle lies to the upper right of the figure. [Reprinted from R. L. Webber and H. Blum, "Angular Invariant in Developing Human Mandibles," *Science* 206 (1979): 689–691.]

many others, like the vestigial organs such as the appendix or the tail, adaptation is only partial?

Again, in the matter of symmetry, for example, how could a creature get from being asymmetrical to being symmetrical? What adaptive value could there be in gradually becoming more symmetrical? Very simple organisms like coelenterates are symmetrical about their centers, and more complex ones such as annelids have axial symmetry. There must have been a time—presumably during the era of the first mollusks—when a plane-symmetrical body plan emerged from a preceding asymmetrical one. This symmetrical plan may have had advantages, probably to do with duplication of many parts of the body. But it is unlikely that this could have come about gradually. It is more probable that two complementary individuals joined to form the symmetrical whole. Among the mollusks there were left- and right-handed varieties, a pair of such organisms constituting a symmetrical whole. It seems that symmetry may have come from a larger unit than the individual. This suggests an interesting speculation—that advanced life forms are essentially two creations in one.

However, there is another way in which symmetry could have occurred—through complementary strands of DNA. It is not hard to see how this complementarity can be exploited to produce two forms, each the mirror image of the other.[29] Once DNA had evolved, symmetry may have been inevitable.

There is also the question raised by regularities in biological structures in general. These regularities are again pervasive throughout the animal and plant kingdoms. For example, the triple angle of the human mandible is a feature of the body that remains remarkably constant between different individuals and across ages as individuals grow up. It has been found by Webber and Blum[30] that the triple angle measured in dental patients at nine different ages has the values 130°, 140°, and 90°, with a standard deviation of around 2 to 3° (see figure 2.1).

This similarity of angle was found to apply even to severely deformed patients, implying a remarkably constant basis in genetics for this geometrically regular feature. In the lung, regular and invariant branching patterns occur in a wide range of mammalian species.[31] In some trees, the angle of emergence of branches from the trunk also shows remarkable constancy between individual plants. West, Brown, and Enquist[32] have derived a universal scaling relationship for these dendritic networks, not only across individuals of the same species but effectively across all species, amounting to what they call a universal scaling law. Although they attribute its existence to the power of natural selection, they point out that this law fits exactly with the fractal considerations of exchange surface areas across such units as leaves, capillaries, and mitochondria: "Fractal-like networks effectively endow Life with an additional fourth spatial dimension. . . . These design principles are independent of detailed dynamics and explicit models and should apply to virtually all organisms" (p. 1677).

These extreme regularities in natural form are very hard for neo-Darwinian theory to account for, but as we shall see in the next chapter, there are other ways of accounting for them that are much more promising. Why should organisms display the regularities and symmetries that we observe? The only plausible way in which such regularities of form can have arisen is through the expression of a process giving rise to a mathematical pattern. Such a process, if it generates certain aspects of the form of every individual member of the species, must be the driving force behind species morphology and therefore, presumably, evolutionary change from one species to another. Shape can only be the product of processes that are ultimately expressible mathematically. It is, in other words, an unavoidable conclusion that many features of morphology are determined by mathematical processes and that evolution results from changes in the values of the parameters of this process, thus giving rise to variations upon which natural selection may act, if these variations are not sufficiently well adapted to the environment. The changes in parameter values will come about by mutation from time to time, producing jumps in morphology and capable of generating wholly new and complex features in one mutational step.

3

THE ORIGIN OF "SPECIES"

Seasons which bring all things. (FRAGMENT 100)

IS IT SPECIES THAT ORIGINATE?

Although in one way Darwin's greatest book was about the idea of species, in another way it was dead against it. Darwin made the idea of species central to his theory because it showed the power of natural selection, which could produce two types of creatures so different that they would be unable to interbreed, yet were both descended from a common stock by a continuous process, the changes having resulted purely and simply from natural selection. On the other hand, *The Origin of Species* is one long argument against the very idea of species: the word *species* is derived from the same root as *special*, and so implies that each is a separate entity, created in some way to be so. This was exactly the medieval notion that the *Origin* was designed, implicitly or explicitly, to overthrow.

The idea of species has been a long-standing problem for biologists, and it is scarcely less so today. No one definition of the word can be agreed on—there are at least four separate definitions of species jostling for primacy. If species is a central idea, it is at present a singularly ill-defined one. The notion of a species is an emergon[1] par excellence: one of the aims of the science of biology should be either to give it a precise definition or else to discard it from its terminology and thinking.

Traditionally there were three different concepts underlying the term *species* and more have been added recently. The *Linnaean concept* of species is one of ideal types, perceptually recognizable as separate and distinct. This is the meaning of the term accepted by most biologists as closest to an intuitive one.

We can "see," as Linnaeus saw, that a specimen belongs to a separate species. But what is seen is not necessarily what is there. We tend to project onto the outer world what are in fact the products of our own perceptual system. Also, the Linnaean classification of species is concerned only with what is visible. For example, it tells us what a specimen is, but not what its descendants may become; genes are a much more recent discovery. It follows that the Linnaean notion of species is one that cannot now stand up by itself as a concept.

A modification of the Linnaean concept is the attempt to classify species in terms of their form and structure, leading to the *morphological concept* of species. This approach resulted from the impact of Darwinian thinking on the idea of the immutability of the essence of species, supposedly specially and distinctly created, which had been accepted prior to evolutionary thought and which underlay the Linnaean classifications. The morphological concept of species is an attempt to make precise the intuitive notion of species on which Linnaean classification depends. However, for the present purpose it is exactly this basis of classification that is being called into question, so this definition will not do either.

The *biological concept* of species is founded on the criterion of what is known as the reproductive isolation of a group. If two individuals can interbreed, then they are considered to be members of the same species; outside this limit there is reproductive isolation. This certainly appears to be an operational definition of something real. But here too there are problems and inconsistencies. Some types of animals that one might want to call different species can nevertheless interbreed, like the canary and the finch. In other cases it may be that only a small proportion of matings will be productive, or that there will be descendants but that some or all of these will be infertile, as with the ass and the mare. It is a convenience to know if any mating will produce offspring or not, but this is not necessarily the criterion we require for species definition.

A more recent variation of the biological concept is what is called the *recognition concept* of species. This would define a species as those individuals recognized by one another as capable of being mated with. One possible link between these concepts is through the approach of evolutionary psychology, in which our perceptual system is supposed to have been shaped by the struggle for survival (in the case of humans, often very recently).

There is another definition of species latent in Darwin's theory, and that is the *ecological concept*. A species according to this definition has evolved to occupy the ecological niche that is available for it. Unfortunately for this concept, it does not uniquely define a species, though it may contribute to its definition. Two distinct species, even of the same order, may occupy the same niche, like the rhinoceros of Africa and southern Asia and the toxodont in

South America. Complete reproductive isolation would seem to rule out a meaningful definition along these lines.

Since the rise of molecular biology the *genetic concept* of species has been added to these by studies of resemblances and commonalties between nucleic acids. These can be used to define a separate type very clearly, but they do not tell us anything about its mutability or otherwise or, more importantly, where to set the type boundaries. There is still no systematic way of classifying genetic differences for this purpose and therefore no readily acceptable definition of species.

THE CASE OF DARWIN'S FINCHES

A theory of evolution must do two things: it must account for variation and it must account for speciation. Natural selection does a good job of accounting for speciation until we begin to ask deep questions. Why do some species apparently progress in a series? Why do they differ, not so much, but so little? For example, a giraffe and a human have the same number of vertebrae. The skeleton of a bird is not much different from that of a horse, except in scale, or that of a shrew from a human being. Even more surprisingly, the embryos of creatures as widely different as fish and humans are said to bear striking similarities in the early stages of their development.[2] Why are these species, if they are the result of variations acted upon by natural selection, basically so similar? Why has nature not been more adventurous in her experiments with form? Why are more diverse and exotic shapes not found?

It might be more accurate to say that natural selection is a good explanation for the *refinement* of species. One way in which this might happen is by geographical separation. This idea suggested itself to Darwin during his visits to groups of neighboring but effectively isolated islands. A subpopulation of a species might be cut off from the rest and, after a long enough period, become a new species.

Darwin's finches are an example of this. These birds of the same genus differ with respect to coloring, the size and shape of their beaks, and other features. They inhabit the Galapagos chain of islands, where Darwin did much of the research that formed the basis of his theory of evolution. The birds differ in several important respects, depending on the exact site that they inhabit. Darwin suggested that the geographical separation was probably the "ultimate cause" of their variation. (Ultimate cause here can be taken to mean distant or original cause, because the immediate cause must be differences in the genetic make-up of the birds, which Darwin could not have known.) They will breed true, although interbreeding will produce intermediate types. They do not, however, constitute members of separate species.

Some of the birds have long pointed beaks; some short, powerful ones. Some of them have brightly colored plumage, while others are more drab. Why this should be so is an excellent illustration of the processes of natural selection. It is of adaptive advantage for some birds to have strong beaks capable of exerting greater leverage for cracking nuts; it is beneficial for others have long pointed beaks that are useful for probing the mud around the edges of waterways.

If all that neo-Darwinism claimed was that Darwin's finches are an example of natural selection, this would be nothing to disagree with. But it is also maintained—it's in the textbooks—that the finches are a piece of contemporary evidence showing evolution taking place, and the implication is that this is the way in which new species develop. What this means is that the selectionist hopes, and believes, that the processes that separated the finches into their different types would, if sufficiently prolonged, one day separate them so far that they would be different species, presumably incapable of interbreeding in the sense of producing fertile offspring. Just when that day would arrive—where that barrier would arise—is not at all certain, nor is the nature of the changes in the finches that would constitute different species.

The size of the upper and lower parts of the beak is different from one type of bird to another, but the topology and morphology of those parts is the same. Coloration varies between the birds, but the shape of the bird is the same. This suggests variations about a central theme, not a change in that theme or a divergence into two separate themes. The time scales involved are also very discrepant. All the Darwin finches are supposed to be descended from a common ancestor of about one million years ago, yet suitable environmental conditions can elicit changes between different varieties of Darwin's finch within a year or so. At such a rate of change one might expect that the finches would have evolved so far from one another in the past million years as to be unrecognizably different, yet they have not.

It must surely be apparent that what is happening here is something different from evolution as it is usually understood. Darwin's finches show variation within a species, and such variation can be expected to occur within nearly all species and at almost all times. If people with heritable traits interbreed, they will come to constitute subpopulations of individuals, with a particular trait common or even universal among them. Human ethnic groups share characteristics such as stature, color of eyes, color of skin, and color and type of hair, whether straight or curly. Groups of people living in sunny and hot areas of the earth have developed plentiful layers of melanin, which protect them from ultraviolet radiation. Conversely, people living in the higher latitudes have developed pale skins, which admit more light and enable an adequate production of vitamin D. This does not make either of these groups into new species, and it takes a lot to believe that any amount of selection along such lines would ever result in their becoming different species. In H. G. Wells' story "*The Time Machine,*" the time traveler finds that in A.D. 802,000 the human race has become divided in two: the industrious, subterranean Morlocks and the effete, surface-dwelling Eloi upon whom they prey. The division had been produced, in a Darwinian manner, by two classes of humans living in different environments, the Morlocks being the descendants of lower-class underground workers. Even allowing for the rather brief time scale, this does not seem a likely scenario.

As the example of Darwin's finches shows, while natural selection does a good job of explaining variation within species, it does not seem to explain the event of speciation itself. It also illustrates the confusion of the concepts of variation and selection, which arises very easily when evolution is being discussed. One cause of variation considered by Darwin was the difference in conditions between separate groups. The difference in the case of Darwin's finches, as in most cases of natural selection, is geographical. But while separation can *allow* variation, it cannot *induce* it. If different groups are subject to different conditions, some members of those groups will survive more readily than others; hence the groups will come to differ from one another. This is natural selection. But to conclude that the different environments actually produce variation is only a step away from thinking that it is selection that produces variation. Nothing could be more wrong than this: selection cannot produce anything; it can only subtract something.

When it comes to the origin of variation, Darwin has only unsatisfactory answers. In *The Origin of Species* he sometimes attributes variations to chance, sometimes to the effects of climate. Sometimes he gives himself over to Lamarkian ideas of the heritability of acquired characteristics and suggests that the life history of the individual actions of group members will determine the make-up of their descendants. He tries one idea after another, without fixing on any one. In the end he has to admit that no satisfactory answer to the question exists, and confesses that "our ignorance of the laws of variation is profound."[3] Indeed, it is not to be wondered that Darwin could not begin to guess at the true mechanism of variation, since it is dependent on alterations in submicroscopic chemical structures whose existence was unsuspected and about whose structure almost nothing was to be known for a century. This is partly why Darwinism was not fully accepted as an orthodox scientific theory until genetics could propose a basis for morphological change.

MUTATIS MUTANDIS

If natural selection does not produce variation, what does? Variation is essentially the same as mutation; indeed, the two are synonymous. The difference is that the term *mutation* is applied to genetic material (genotype) while variation applies to expressed characteristics (phenotype). To inquire about the sources of variation is to ask what mutations are and what produces them. Mutations can affect the DNA of the body cells (somatic mutations) or of the gametes (genetic mutations) and it is the latter that are of importance for evolution.

Mutants, both natural and artificial, have been intensively used as experimental tools in the laboratory in the last decade, but the mechanisms by which mutations result in different expressed characteristics are still obscure. It is well established that gene changes are expressed in conjunction with environmental influences in the bodily form of the organism-genotype gives rise to phenotype. It is also accepted that the reverse cannot happen—phenotype cannot directly affect genotype. No matter what behavior you practice, it cannot directly alter your unconcieved children's genes.[4] However, the genes can be altered by a number of other factors, such as radiation, viruses, trauma, and chemical agents.

Mutations can be produced artificially, though not always predictably, in the laboratory by genetic engineering. There are also spontaneous mutations, that is, mutations for which there is no known cause. These must be the most interesting ones from the point of view of evolution. They are the raw material on which the mill of natural selection can grind, so it is surely important for us to know as much about them as possible. What produces them, and what kind of variations result? Do mutations emerge in a random or a systematic fashion?

It is often stated as an article of selectionist faith that mutation is much less powerful as a source of variation than is selection, that is, environmental factors. But that is nonsensical, since both are essential for any change. To make selection account entirely for change is stretching one answer to cover two questions: first, how do heritable changes in organisms occur; and second, why do some changes get passed on while others do not? The selection principle can answer only the second question, but an attempt is made by some neo-Darwinians to make it answer the first as well.

The idea of species might be of more use to a selectionist if distinct species could be associated with distinct environmental conditions. But this is not the case: types that are descended from one another are to be found in the same geographical location, particularly in the case of plants, while the same species are found distributed in widely differing environments.[5] It is an inescapable conclusion that the gross features of the environment are not a very important factor most of the time. There are only further problems to be found in trying to associate the idea of species with a particular environment.

The concept of species has also been found to present problems to a hypothesis of evolution by small mutations. A small mutation is usually not sufficient to produce a subspecies, much less a species. An accumulation of mutations amounting to something like reproductive isolation would be required for speciation, but as we shall see later, mutations need not be small. The picture of speciation according to the orthodox view is that many individuals of a species acquire small changes until they collectively form a subspecies that is better adapted. But how does it happen that the favorable mutation occurs in enough individuals at the same time? If it does not, wouldn't interbreeding between mutants and nonmutants dilute the effects of the mutation?

THEORIES COME AND GO: THE PEPPERED MOTH REMAINS

So we are driven to ask whether there really are such things as species. Is it just another case of looking for "atoms" that, if only they could be found, could be assumed to be the building blocks of biology, but are as illusory as the atoms in many other spheres of science? Do we need the concept of species at all, and if so, what for? The idea of species is an awkward one for biology in general, as we have seen, but it presents particular difficulties for Darwinism. The problem for the Darwinian is that the whole concept of species is really at odds with the idea of gradual change upon which evolution by natural selection depends. If species can change into other species by gradual steps, where do you draw the line between them, and, indeed, why have a division at all? Darwinian explanations imply gradual change, in the sense of small changes to an organism, or changes to only one part of an organism at any one time, or both.

Unfortunately for the Darwinian case, such gradual changes occur observably only within closely allied organisms, usually members of the same species. Thus Darwin's finches vary in bill morphology, coloration, and so on, but are all accepted as members the genus *Geospiza*. The peppered moth, *Biston betularia,* which inhabits the north of England, exhibited markedly darker coloration as the Industrial Revolution progressed, caused by higher rates of predation by birds of the light colored moths, more clearly visible against the now-blackened trees. This is one of the most frequently cited examples of natural selection in action and it has been much disputed,[6] but even if the changes were genuine, there is no suggestion that the moths displaying this melanism were of a different species. With the air now free of tree-darkening pollutants, the moths are reverting to their former coloration.[7] The conclusion would seem to be that natural selection can certainly pick out desirable traits from a pool of such variable characteristics, but this process will not of itself lead to speciation.

It appears that evolution does sometimes proceed by jumps and perhaps only by jumps, and that the genetic variations producing speciation events must therefore also be occurring in jumps. In contrast to classical Darwinism, the so-called punctuated equilibrium theory due to Gould and Eldredge[8] proposes that evolution exhibits short periods of rapid change punctuating long intervening periods of stability. With such a process there would not be a large number of intermediate forms between distinct species. This theory is supported by the fossil record, which usually does not contain such intermediates. The explanation put forward by Darwinian theory to account for these sudden jumps is that many small genetic changes of organisms can take place that are due to natural selection but are not visible in morphology, lying dormant until they are expressed at a later time as changes that are observable. The problem with this explanation is that it is hard to see how natural selection can act on characteristics that do not result in visible morphological change. The argument that accumulations of unexpressed variations arise by natural selection is not convincing as a general mechanism of speciation, although it might apply in particular cases. How could a number of neutral mutations—neutral, that is, from the survival point of view—accumulate without expression, somehow to amount to an adaptive change when they are expressed? Only by chance, but chance is not a very satisfactory explanation of events. Punctuated equilibrium is a model much more easily explained under a directed-mutation hypothesis of evolution than under a natural-selection model. The mutation hypothesis predicts that changes in expressed characteristics will occur from time to time because of changes in the genetic material. It also implies a stable state in between these changes. In support of this view, it has been found that there is a good correlation between the times of occurrence of changes in regulatory genetic material and the times of changes in morphology.[9]

Mutations are sometimes classified into two types: those for which a cause is known and those for which no cause is known, the latter being called spontaneous mutations. The cause of a mutation is known as a mutagen. A typical mutagen is something like X rays, ultraviolet light, or a chemical agent. The mutagen may cause a change to a point (locus) within a gene or it may cause an alteration to a chromosome, inserting, deleting, or altering a longer sequence—a chromosomal segment. Mutations of the latter type are the most severe, probably causing gross changes to anatomy and function, and these are the ones that are thought to drive evolution forward.

There is an apparent contradiction here, since, if mutations are for the most part deleterious, how can they lead to improved species? The answer is that most mutations do result in deterioration and are usually lethal, but every now and then a mutation that is nonlethal, neutral, or even adaptive may occur. Of these, many will be nonheritable and will disappear in the next generation, but a very few will be heritable and so may lead to a new species being established. Since the frequency of adaptive, heritable mutations is very low, the formation of a new species will be a very rare event.

Thus under a mutation-driven view of evolution, there would be general stability, but from time to time new types of organisms would appear, most of them minor variations but some of them dramatically different. Most of these new individuals would not survive long enough to reproduce, but some would survive. Since their genes have been altered, they pass on the mutation to their descendants, and in this way a new species emerges.

HOW SYSTEMS SELF-ORGANIZE

It is one thing to assume that mutational variation will explain the observed rate of evolution and another to show how it does. If we assume that mutations are the agents of evolutionary change, the question becomes how to account for mutations of an adequate number and variety to bring about the observed rate of evolution. The difficulty is that, if mutations are unsystematic, how can they have resulted in such a well-patterned outcome? The legendary monkeys at their word processors could not write the works of Shakespeare just by chance. In the history of the world they would not complete even a single line.

Mutation rates would have had to be directed in some way in order to produce creatures like those presently observed. On the face of it, there are far too many possibilities that might arise in unpatterned mutation. Ulam[10] has calculated that, if achieving a significant advantage, such as the human visual system, requires 10^6 changes, then it will take 10^{13} generations in a population of 10^{11} individuals for the change to become established. If there is one generation per day, this means several billion years. Other attempts to calculate the rate at which changes need to occur for mutations by themselves to become effective have also yielded enormously long times. Chance mutations will not suffice; mutation must be in some way controlled, directed, or patterned if it is to do the job required of it.

The problem for mutation theories is then to explain how mutations are organized in order to arrive at a goal in a reasonable time. This process of organization would be the analog of the genetic algorithm, in which the goal is known or is at the end of a directed path. Answers are beginning to emerge: for example, Stuart Kauffman[11] has suggested that genes form a self-organizing system. He also introduces what he calls a "fitness landscape" in which these groups of genes are subject to selective pressure and the solution moves "uphill" to an optimum value. This raises an interesting problem: mountain tops are not next to one another; you have to go downhill to reach the next one. In evolutionary terms this means that in order to evolve further, the organism would have to become *less* fit.

The classical question posed to Darwinians is: What use is half an eye? If the eye had emerged as the result of single, separate mutations giving rise to gradual approaches to its shape and function, might not some of these intermediate stages have been positively disadvantageous? It is certainly likely that some would have been useless from the point of view of survival. In either case, individuals with such mutations would not have been selected and the eye as a useful organ would never have evolved. But this difficulty is equally strong for mutation-based theories of evolution, which assume many "random" mutations. Even if the emergent organ were never subjected to any selective pressure, it would not advance in the time available for evolution to the stage where it became a usable feature. But if we have, instead of many unrelated point mutations, a series of directed steps, then the time scale does become feasible. If mutations occur in a patterned way, some advantageous mutations must emerge as a result of the directed process that would never have time to occur if the mutations occurred randomly.

Another factor, not much emphasized, yet crucial, and one that it is difficult to explain under any but a directed-mutation hypothesis, is that species almost invariably evolve in the direction of increasing complexity. Evolution seems always, or nearly always, to tend toward the more complicated and elaborate form. This is so fundamental a fact about life that it is almost a definition of evolution for many people, and it precedes any notions of how that evolution came about. Henri Bergson made it the basis of his ideas, in which evolution is the expression of the "life-force" striving ever on and upward. It is certainly true that this drift toward complexity is the general tendency of evolution (though there are also many examples of "backward evolution" in which simpler organisms appear as successors to more complex ones) and it is a tendency that is very hard to explain either by undirected mutation or on the basis of natural selection. Why should the net direction of adaptations selected be toward complexity? Why should it be a general principle that a more complex organism is better adapted than a simpler one?

PRUNING THE TREE OF LIFE

If the mechanism that had been the driving force in evolution had been entirely undirected, then we might have expected a rather different kind of life on earth. Natural selection would have to act at each step on organisms that were of a certain level of complexity. Mutation within a species will produce individuals, some of which are more complex and some less complex than their progenitors. If natural selection kills off the less well adapted of these, leaving the others, what reason is there to expect that the remainder will be more or less complex than before?

Evolution of this type could be modeled as a kind of Markov chain. A Markov chain is a sequence of values, each of which depends on the one before it. Markov chains may be constructed that grow to a certain degree of complexity and then collapse. A good example of this kind of Markov chain is a gambler's wealth as a result of a sequence of bets on the red or black in roulette. If you bet on these two colors only, assuming they are exclusive possibilities, your winnings may grow for a while, but they will decrease again and you will eventually lose everything you have won, including the initial stake. If you then start again with a new stake, you will get a new sequence of wins and losses. The length of the sequence before the inevitable happens varies: some sequences of wins and losses will be very short, some longer, but not many will be very long.

It would seem that variation in a new organism is as likely to be in one direction as another; that is, as likely to lead to more complexity as to less and as likely to lead to increased as to reduced viability. So, had natural selection been the only force acting to produce change, rather than the few basic patterns for living things that we find now, we might have expected to find many more. But this is not what is found: instead of there being a wide variety of different shapes of living things, in advanced organisms at least we find only one basic shape—the tree-like or dendritic shape. Most eukaryotic life uses this basic pattern, which is thus a good candidate for the universal body-plan of advanced life. The body is composed of a central trunk with roughly symmetrically branches, the branches subdividing repeatedly. In animals the subdivision of the body ends after three or four repetitions; in trees it may be repeated more than half a dozen times. The two halves of the animal body are highly symmetrical, although the asymmetries may be significant; for example, the suggestion that hemispheric specialization in the human brain has consequences for psychological functioning.

Not only is the dendritic form a plan of the soma, or body, of the organism, it is also found in the shape of the subunits of which the body is composed. In man and the higher vertebrates the nervous, circulatory, and pulmonary systems all follow the same plan as the body itself; that is, they are dendritic. This shape is a roughly tree-like structure, but a tree that has been folded into a hypothetical fourth dimension so that "roots" and "branches"—afferent and efferent neurons, veins and arteries—have been brought into juxtaposition. A tree transformed in this way would have as its folded "trunk" a bulbous end with a corresponding grooved crown such as that found in the heart and brain.

So striking is this structural resemblance that it might be convincingly said that a human being is three trees—the tree of the trunk, the tree of the nervous system, and the tree of the vascular system. These features of higher plant and animal morphology—dendritic self-similarity and symmetry—both demand explanation and are at the same time a powerful clue to their origins.

To summarize the objections to neo-Darwinism:

1. Long though they are, time scales are too short to accommodate the changes required.
2. Evolution takes place, not gradually, but in jumps (Gould & Eldredge).
3. The distribution of species under neo-Darwinism should be different from what is observed.
4. Neo-Darwinism postulates stochastic processes leading to exotic variations. If this were correct, species should be simpler and more varied, like those produced by Conway's *Game of Life*.[12]
5. There is no attempt in the neo-Darwinian theory to explain the ever-increasing complexity of living things.
6. Neo-Darwinism fails to account for the regularities and universality of form in advanced living things.

MOTHS COME AND GO, BUT THE THEORY REMAINS

We have now seen how and why the inappropriate application of a random-selection model to biology is failing. Darwinism and molecular biology represent opposite poles of an attempt to come to grips with the subject matter of biology at a level other than that of biology itself. Both simplify, and both oust the study of biology from its own field. The Darwinian theory survives. Chemical thinking rules. Such an approach works when we are not dealing with systems or with only very simple ones, but it will not work with complex systems.

Complex systems differ from simple ones. A complex system can be understood only in terms of its own laws of functioning, because these laws exist at no other level and in no other discipline. Organisms are highly complex systems,, and if we are to come to grips with them, we must have theories that deal with such systems explicitly. In the case of biology in particular, it is no use expecting that chemistry and evolutionary theory between them will somehow fill the gap left by our lack of understanding about how organisms work.

How might we begin to go about getting such an understanding? One of the central themes of this book is that complex systems, including those of living things, are to be understood in terms of a mathematical theory. Such an approach is anathema to the traditional neo-Darwinian, who is wedded to the concept of randomness and who thinks directedness is next to godliness. But I ask such thinkers to set aside their preconceptions and consider the evidence in an objective spirit. What is so different about evolutionary processes that they should not also have underlying laws? If it is acceptable that physical, chemical, and biological processes should proceed in a directed way according to natural laws, why should evolution not follow a similar path? I shall try to demonstrate in chapter 6 what this process may be and that it is describable in terms of the mathematics of iteration. For now, I would like to make the point that it is not out of the question that evolution may be a directed process, since direction is implied by all fully developed scientific laws.

The overall tendency of such mathematically driven systems is toward the more complex, though occasional backward movements may be expected. All such systems are based on iterative processes that are of universal occurrence. Organisms are iterative systems by their very nature: the division of cells, the consequent growth and development of the organism, the processes of feeding and excreting and reproducing, all these are iterative. Not only do organisms have this iterative character, but so does their environment: day follows night, ebb tide follows flood tide, and year follows year. As it appears, the very nature of time itself is iterative—the word we use for time is cognate with that for tide. Time and tide are essentially the succession of cycles, and in all devices that measure time, cycles are produced by systems having a periodic orbit. Time is iteration, and in this world there is no escape from iteration or from time.

4

CHAOS AND DIMENSIONALITY

Even the barley-drink[1] disintegrates if it is not moved.
(FRAGMENT 125)

THE NATURE OF CHAOS

As we discussed in chapter 1, iteration is the most basic process that can be found in the world. Now I want to explain how this process is expressed. Two of the phenomena in which the iterative process is most clearly marked are the operation of deterministic chaos and the production of fractal geometry. Fractals and chaos are intimately linked: chaotic systems are controlled by iterative processes, and fractal forms are the outcome of iteration. Together these are the two most important keys to unlocking the iterative processes of nature. They are also essential tools in removing some of the obstacles that have arisen in natural science.

Scientists, like other creative people, need mental models in order to think about their subjects. The picture of the world that most scientists have in mind is Newtonian. In this view, determinism and order apply locally but there is disorder and uncertainty on a large scale. For example, we may understand the way two colliding molecules interact, but it is hard to understand the behavior of a gas; or we may know the laws of gravitation but be unable to see any pattern in a cluster of stars. The Newtonian worldview assumes that everything can be understood in its terms and that even the large-scale events could be understood if more information were available: life, the biggest puzzle of all, is filled with confusing detail, but further knowledge of the details should reduce it to order. According to this view, order will eventually become apparent, and disorder will be rooted out.

When we examine this model further, however, matters turn out to be very different from what we expect: chaos is in general a better model of the world than order; events are for the most part even less predictable on a small scale than on a large scale. Far from being incidental to living things, chaos is both natural and beneficial. In this new worldview, chaos is the most valuable attribute that any system can possess, whether it is a living organism, a human society, an economy, or the ecology of a planet.

To understand why this is so, and to understand the necessary and positive nature of chaos, we must first see how chaos theory came to be discovered and what led to its development, from the first intimations that nature might be not as orderly as was hoped, to the current state of things, in which we are finding evidence of chaos everywhere.

THE NEWTONIAN WORLD-PICTURE

For the last three centuries science has been dominated by the Newtonian model, that is, the worldview arising out of the work of Sir Isaac Newton in the late seventeenth and early eighteenth centuries. Following Newton, a standard physical model of the world emerged: one of particles (objects that had mass but no size) attracted to one another by gravitation and moving according to the laws that Newton, among others, had described.

This model fitted well with the mathematical analysis called the calculus.[2] Calculus was applied early to the problems of motion of idealized particles in space (though Newton would not have put it quite like that). There are two aspects to the calculus: one is concerned with cutting things up (conceptually) into little fragments (differential calculus), and the other, inverse, method, with assembling them into wholes (integral calculus). The differential and integral calculus is one of the most powerful tools ever devised and, as can be seen, it is basically iterative: Newton did in fact use iterative methods extensively to solve equations.[3]

At the same time as Newton, Gottfreid Leibniz was also working with calculus, but using a rather different method. Newton's method used what he called fluxions (from a word meaning "a flow"), while Leibniz used infinitesimals—very small units that were not zero but could be made as small as desired and therefore approached zero. The differences between Newton's and Leibniz' methods, while at first sight largely notational, are significant, because fluxions led to the idea of the rate of change of a function, while infinitesimals were debatable units, no one being sure whether or not they could be properly defined, even until the present.[4] As far as practical use of the notation went, however, Leibniz' was the more useful, both in calculations and in working with transformed formulas.

Strangely, Newton and Leibniz espoused two alternative and rival notations for the calculus that were almost exactly opposite to the viewpoints of their philosophical positions. Newton believed (at least for analytical purposes) in particles, which are discrete but bound together by forces that are continuous, while Leibniz believed in holistic units called monads, each mirroring all the others, rather like nineteen-seventies-style architecture in which each building is covered in reflecting surfaces, enabling it to parasitize its surroundings.

In the Newtonian world-model, the world was viewed as essentially deterministic (it obeyed laws). Determinism gave rise to the idea of predictability; that is, the future state of the world could be predicted from its present state, provided we know the laws of motion governing the bodies that comprise it and their initial conditions. This led to the belief that the future state of a mechanical system could, in principle, be forecast as certainly as predicting the future movement of the hands of a clock, given the present time. This Newtonian worldview was summed up in its most complete form in the eighteenth century by Laplace, who believed not only that the motions of the celestial bodies of the solar system were predictable, but that all other systems, being composed of particles and functioning according to known laws, were, in principle, also predictable. If the world consists of particles, and if these can be described in terms of a few laws, it ought to be possible to discover the principles governing any system in the world and thus predict, understand, and perhaps ultimately control it.

THE DETERMINISTIC WORLD

This deterministic view came to be accepted by most scientists. It was still accepted at the beginning of the twentieth century when Bertrand Russell[5] only half-jokingly said the day was not far off when, if you wanted to know the answer to a question concerning the future, you could easily find it by turning the handle of a machine kept in Somerset House, then the home of the British Public Records Office. This model of determinism, hand-in-hand with predictability, was a mental as well as a mathematical one. It evoked in the mind of the scientist a picture of particles orbiting one another in otherwise empty space and in a completely deterministic way—a kind of solar system model, which was applied to every system from galaxies to atomic structure. That picture has lasted until quite recently and is perhaps the one that many scientists, particularly if they were educated before the nineteen eighties, still have in mind.

In fact, from the philosophical point of view, the Newtonian model raises many more questions than it answers. First, there is the question of the particles. How is it possible for something to have mass but no size; that is, for something to be effectively nothing? Then what about the space in which they move? What is in between the particles? Since no one could find any medium that fills the gaps between objects, the answer seemed to be empty space. This empty space has been aptly named container space, because we can picture it containing particles just as a shopping bag holds grocery items. If the particles in container space move according to laws conditioned by their mutual presence, they influence one another by their mere existence in space. But this gives rise to another problem. If the space is truly empty, how can the particles influence one another? This was recognized as a difficulty from the start, and it has only recently been resolved with the notion of an "active" vacuum. Despite these problems, Newton's model worked well in practice, especially in the area of mechanics.

But many kinds of systems remained impossible to predict, and remain so to this day, for a variety of reasons: because of the complexity of the system concerned, or the number of elements involved in it, or a lack of knowledge about—and perhaps inability to find out—where those elements were at a given time. There are many systems like this. Everyday examples are economic trends, human behavior, and the weather. This was a continual embarrassment for pure and applied sciences, but most especially for mechanics. Sometimes the inability to predict was explained in terms of the complexity of the system in question. In psychology, for example, the complexity of the human brain is such that understanding it was thought to require the accumulation of vast amounts of data so that the problem, rather than being one of predictability, was simply one of making enough observations. Unfortunately, even some simple systems turned out to be impossible to predict. Engineers, for example, often worked as much by rule of thumb as by Newtonian mechanics, something very puzzling for the theory of determinism. If a simple system with very few parts and every opportunity to observe them was unpredictable, the reason could hardly lie in the number or complexity of the parts or the lack of knowledge about them.

THE THREE-BODY PROBLEM

The beginnings of an answer came with an insight of the French mathematician Jules Henri Poincaré at the end of the nineteenth century. It led him to suspect the existence of what we now call chaos, although it is not until the last two decades that the topic has begun to arouse interest among scientists in general. In 1887 King Oscar II of Sweden offered a prize for a solution to the question of whether the solar system was or was not stable. Poincaré worked on this problem and found it was very difficult. Since all the bodies that comprise the solar system attract one another, the answer depended on predicting the orbits of several mutually attracting bodies. It proved impossible to predict the orbit of even three interacting bodies very far into the future, because small changes in the initial conditions of the system led to large differences in its final state. In 1890 Poincaré published a paper on the problem of three orbiting bodies and showed that it could not, in general, be solved. This insolubility of the three-body problem, as it came to be known, made the idea of predicting a whole world of particles absurd, except in the sense of making statistical generalizations. Though it was not seen at the time, this realization opened the way for a new worldview. It became clear that the vision of Laplace was incapable of realization, even though it was not yet fully understood why, or what the implications of this were.

The discovery of what we now call chaos was gradual and, as so often in science, accidental or even serendipitous. A large contribution to the science of chaos was made in the nineteen sixties by Edward Lorenz, a meteorologist. Lorenz, who was studying weather forecasting, had modeled the weather system in terms of three simplified equations.[6] Using a computer and starting from a particular set of conditions, he made predictions according to the model and then observed the outcome. On one occasion, in order to save time, he repeated a set of predictions for the same model with the same starting point, but he made approximations in the initial values.[7] Lorenz found to his astonishment that the computer predictions for the second, approximated run turned out very differently from those of the first. The predictions started out the same, but as time passed they diverged more and more, until, although they had similarities of pattern, they were following a completely different trajectory.

What had happened with this particular model was that it had turned out to be highly sensitive to the initial conditions chosen. Varying these even by altering the third decimal place made a great deal of difference to the outcome. It would not have happened with all such models, however. Systems that exhibit chaotic behavior are of the kind known as nonlinear—that is, there is no simple arithmetic relationship between a chosen variable of the system and time. Sensitivity to initial conditions, occurring in nonlinear systems, is now seen as the hallmark of chaos.[8]

This chaotic behavior of a set of nonlinear equations iterated on a computer was probably the first demonstration of the phenomenon of chaos applied to a real-world situation. It had been known for a long time that the outcome of calculations can differ, depending on small changes in initial values. The problem of "ill-conditioned" equations[9] in computing was known to Newton and was encountered later by Poincaré in his attempt to solve the three-body problem. In this sense chaos was not new in computing, even when there were no electronic computers. What made this discovery of Lorenz's important was that the finding was not simply limited to the computer. If laws were applied and seemed to work in the real world, and if the same uncertainty in initial conditions also held, then chaos would be a characteristic of the real world too, at least insofar as we can describe and measure it.

The Lorenz model *did* correspond to the real world in important ways. The weather is certainly difficult to forecast; this *does* seem to be linked to the fact that it is sensitive to small perturbations; the weather *does* move between different states much as the shape of the Lorenz equations suggests; and the use of chaos theory *has* had considerable success in meteorology, enabling forecasts to be checked for reliability and setting time limits on forecasting accuracy.

Since then it has become clear that chaos also appears in many other types of systems, so many in fact as to be a truly universal phenomenon. *Chaos* was the name chosen to describe the way such systems work. It was an unfortunate choice because before that time the idea of chaos was of an utterly confused and unforeseeable set of events, while chaos in the new sense arises out of purely deterministic conditions. Furthermore, the study of chaotic systems has revealed that hidden within chaos there is order of a hitherto unsuspected kind—but a dynamic and frequently unstable order, of much more complexity and universal importance than the simple, static order of Laplace.

CHAOS IN A MIXING BOWL

To see what a chaotic system is like, consider what happens when you roll out and fold a lump of dough. When you stretch and fold a substance, the distance between any two points grows in a nonlinear fashion. If the points are separated by a certain distance, then each time you stretch, you double that distance. However, you must also allow for the folding process, which brings some of the points closer together again when they have crossed the half-way point.

When you repeatedly roll out and fold dough, the distance between any two given particles of dough approximately doubles each time. That means that the error of measurement of the position of a particle also doubles, but not exactly. If we know the point's position with a certain precision to start with, we will know it with only half that precision after the first stretch and fold, and this loss of precision will double each time we carry out the operation. After many such folding and stretching operations we will have lost all the precision we had started with. A particle embedded in the dough whose initial position is known as accurately as possible (such as a cake decoration) will move about in an increasingly unpredictable way, so that after a few folds it will be completely lost. This is an easy experiment to try for yourself.

The certainty is lost because the doubling of distances amplifies the initial error in measurement. If we imagine that the folding takes very little time and the stretching is continuous, then the loss of certainty is also continuous. As we lose certainty, we also add complexity.[10] In many situations matter is being stirred and folded continuously, like a cloud of gas or swirling water, so this is a very general model of dynamical situations. A good example is the mixing of substances in a domestic food mixer. In a food mixer, the stirring, and hence the stretching and folding, goes on continuously. After a few rotations such a system becomes so complex that you can no longer describe it in terms of its initial conditions and a few simple laws. In an experimental physical system, unlike the food-mixing process, we want to keep track of it and, if possible, forecast its future state, but we can only do this within certain limits. Those limits are constantly widening as more and more complexity is added to the system. In time, so much complexity is added that the future position of a particle becomes completely unknowable. Thus an apparently infinite degree of complexity can be produced by the repetition of a simple process. At the same time that information is being lost, it is also being added. The process adds complexity information but loses positional information. The two exactly balance, so that there is no net information change over time.

From observations such as these a great deal follows that is upsetting to the traditional models of science. If the solar system is more like a food mixer than a clock, then we require new models for thinking about physics. Mental models, as has been said, are of considerable importance to the scientist, particularly the theoretical scientist, who functions in many ways like a creative person in any other field. When the mental models traditionally used in a science are found to be inappropriate, it amounts to a crisis in the theory of that science. If they are models that cover a whole range of sciences, then the crisis is even more profound. If simple deterministic models will not suffice even for the hard sciences, they will certainly not provide useful insights into the so-called soft ones, such as biology or human behavior.

THE UNIVERSALITY OF CHAOS

Many systems in the real world seem to be of this type. Contrary to the orderliness that Laplace expected, they display a great variety and unpredictability, and even the apparently orderly systems of planetary motion and mechanical movement are really chaotic when seen close up. More obviously chaotic are water boiling in a kettle or chemicals reacting in a tank. The mathematics that has developed out of chaos theory has enabled scientists to describe such systems better. All of them share the same characteristics and this has led to the idea of *universality in chaos*[11]—the idea that the same laws apply to a wide variety of systems. The classic chaotic system is fluid flow. It would not be far wrong to picture all chaotic systems in this way: the waterfall and the boiling kettle are the chaologist's replacement models for the orbiting particles of Newton and Laplace.

From the point of view of chaos theory, the flaw in the Laplacian model is not determinism—chaotic systems may in theory be completely deterministic—it is predictability. Even deterministic systems may not be predictable very far into the future, because some error must always arise in measuring the initial state of a system and, if the system is following nonlinear laws, then that error increases exponentially. To look at it another way, since information is the same thing as uncertainty, then rather than saying that chaos produces loss of certainty, we could just as well say that chaos produces information. A chaotic system is constantly generating new information that could not have been known at the start. The amount of new information is perpetually increasing; that is why we inevitably lose track.

This is one of the primary discoveries of chaos theory and it is important because it undermines part of the worldview that scientists have traditionally used. It raises new and puzzling issues: what happens to the "new" information that is constantly being generated? If we equate information content with entropy, then entropy is increasing, everywhere and always, at an astonishing rate! We shall return to these questions in chapter 9, but the operation of chaos certainly means that most systems are not, in practice, predictable very far into the future. We would do well to acknowledge this fact and not hope, as was done for so long, that the problem will finally be cleared up by more precise observations or some advance in the formulation of the rules governing the system that will lead us to certainty.

ORDER WITHIN CHAOS

The other main surprise to come out of chaos theory was the discovery of the order hidden within many nonlinear systems. At first sight, chaotic motion might seem to be the complete disorder that the term traditionally implied. But such systems, although their behavior seems complicated, have in them an underlying pattern that can be used to predict and sometimes even control them.

The prototype of chaos is the dynamical system, that is, a system governed by deterministic laws that prescribe the way in which it changes over time. Many natural processes are dynamical systems, including, of course, mechanical ones but not restricted to them. The central nervous system—the brain and spinal cord—is a dynamical system, and so, too, is the movement of a football crowd. Associated with any dynamical system, chaotic or not, is a path that describes a variable of that system as it evolves.

For example, a simple pendulum swings to and fro in its path in a regular fashion. At each end of the path the pendulum bob comes to rest, so its speed varies from zero to a maximum and back to zero twice in each complete swing. The position of the bob, as measured by the angle of the string to the vertical, varies from plus θ to minus θ. If we measure the position of the pendulum bob at different times, we have values that can be plotted on a two-dimensional graph and the way the position changes over time can be represented as a diagram (figure 4.1) showing velocity[12] plotted against angle.

If we pick the same starting point for our measurements of angle and velocity, the trajectory traces out a very simple path—a circle. If we measure angle and velocity from different relative positions of the pendulum cycle (called the phase angle), we get one of a number of shapes known as Lissajous figures (see figure 4.2). These shapes show how, for different phase angles, the variables of velocity and position change together in an imaginary space called phase space.[13] A picture like figure 4.1, which plots one variable against another, in this case position against velocity, is called a phase portrait. We need not be restricted by real space any more in visualizing the pendulum. Phase space is not the same as the space we see with our eyes: it is a mathematical construction.

Many dynamical systems are governed by *attractors*. An attractor pulls the trajectory of a variable toward it and eventually claims it. In the case of the simple pendulum, the circular orbit of figure 4.1 is an attractor. If we drop the pendulum bob from our hand, it will follow an erratic path to begin with but it will soon settle into the to and fro movement represented by the circular or Lissajous attractor. We can picture this, rather animistically, as

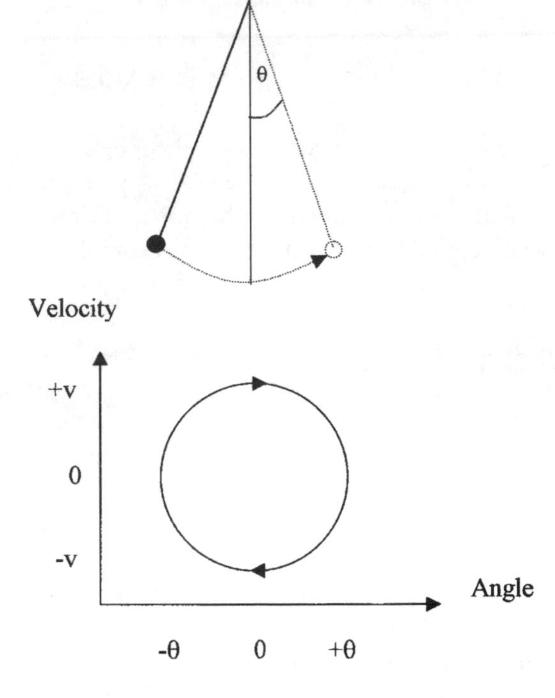

FIGURE 4.1 Motion of a simple pendulum.

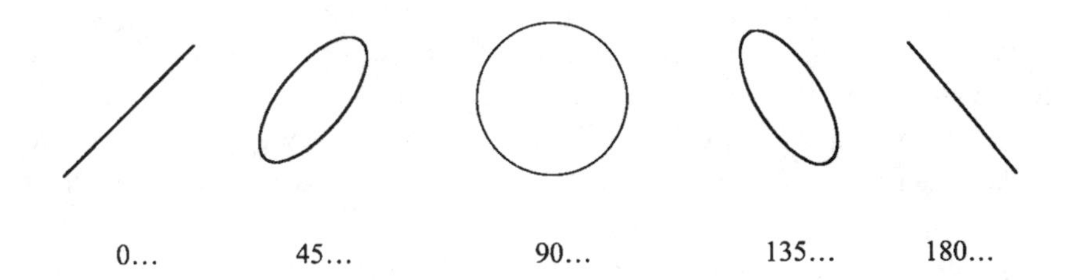

FIGURE 4.2 Lissajous figures for five different phase angles between 0 and 180 degrees.

though the attractor in phase space pulls the orbit of the plumb bob toward it and holds it there.

MAPS OF CHAOS

Another way of representing the dynamics of a system is to plot the same variable at different times. Some work done by Totterdell and his colleagues[14] studied mood changes in office workers. Every two hours they asked workers to rate how much they had felt a particular mood, such as anger or tension. Figure 4.3 shows a diagram called a *Poincaré plot* for three variables: anger, tension, and calm. As can be seen, anger varies considerably between one epoch and the next while tension and calm tend to be stable over a long periods.

Figure 4.4 shows three-dimensional phase portraits of the action of the heart. On one axis the original signal is shown; on the next axis is the signal delayed by a constant amount; and so on. The left-hand figure uses a delay of 12 ms and the right-hand figure, 1200 ms. The two shapes, while different, are related, and particular features of the heart's action, such as the strong P wave, can be most clearly seen on the left-hand diagram.

Yet another way of showing underlying system dynamics is the *return map*. This is a plot of a parameter of the system when another parameter, possibly derived from it, was last at its current value. For example, we could show when the angular velocity of a pendulum last took its current value, using the phase angle on the two occasions as the axis value. This is often useful when looking for periodicity in the system, as such periodic cycles will show up as symmetries in the return map. Figure 4.5 shows the Poincaré section (left) and the return map (right) for normal heart activity as displayed in figure 4.4.[15] The left-hand display is in effect a cross section of the three-dimensional (left-hand) figure 4.4, showing the values when the trajectory cuts through a plane. The Poincaré section shows a clustered formation, which suggests that an attractor may be present. The return map is highly symmetrical about the line $x = y$, showing the presence of strong periodicity in the data (and therefore in the rhythmical activity of the heart).

In a kettle of water being brought to the boil, the motion of an individual molecule of water has a much less simple trajectory. It too has a corresponding trajectory in phase space but, although it is not a simple Euclidean shape, neither is it amorphous. It displays a recognizable form and has regular features. Surprisingly, systems like this are also governed by attractors, but in the case of chaotic systems these attractors are much more complicated shapes than the circular orbit of the pendulum. If the system is completely predictable, as is true for a lin-

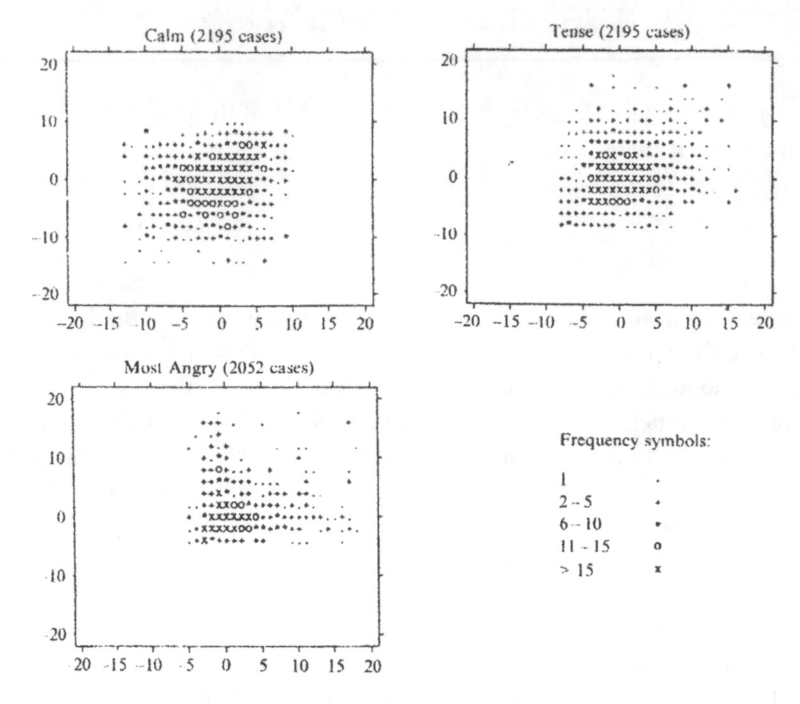

FIGURE 4.3 Poincaré plots of mood changes for calm, tense, and most angry. [From P. Totterdell, R. B. Briner, B. Parkinson, & S. Reynolds, "Fingerprinting Time Series: Dynamic Patterns in Self-Report and Performance Measures Uncovered by a Graphical Nonlinear Method" *British Journal of Psychology*, 87 (1996): 43–60.]

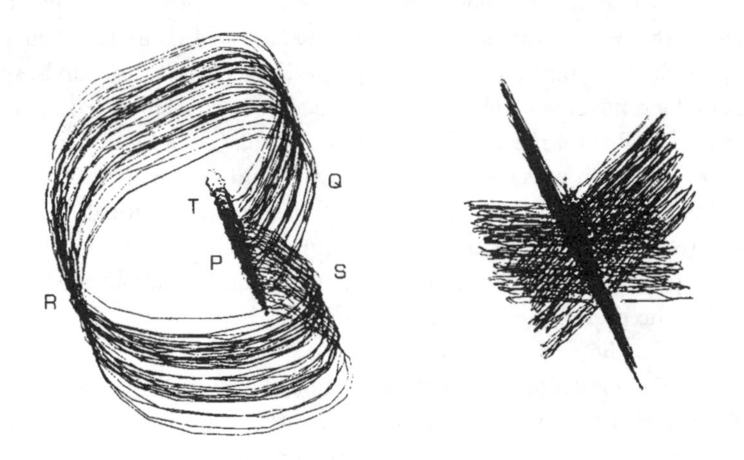

FIGURE 4.4 Phase portraits of heart action for two different tau values. P, Q, R, S, and T represent the principal components of the EKG. [Reprinted from A. Babloyantz and A. Destexhe, "Is the Normal Heart a Periodic Oscillator?" *Biological Cybernetics* 58 (1987): 203–211.]

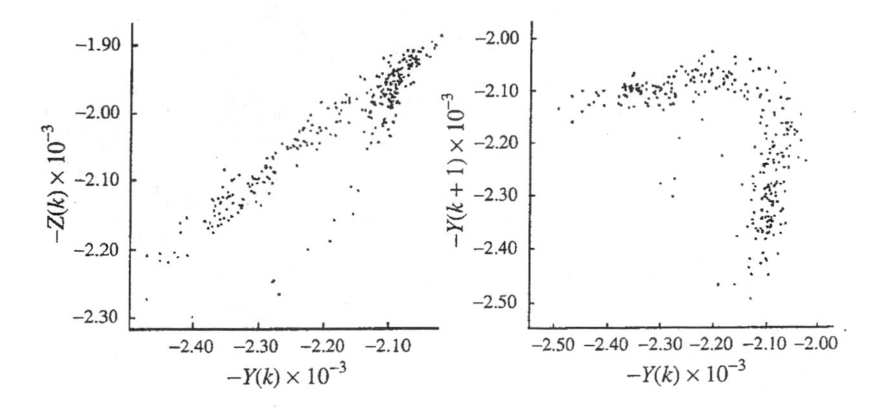

FIGURE 4.5 Poincaré section (a) and return map (b) for heart action. [Reprinted from A. Babloyantz and A. Destexhe, "Is the Normal Heart a Periodic Oscillator?" *Biological Cybernetics* 58 (1987): 203–211.]

ear system, then the attractor will be a simple shape, but if the system is nonlinear, then the shape of the attractor is much more complex. In the case of chaotic systems the form of the attractor is described as *strange*, and this name was picked precisely because the shape was not a straightforward Euclidean[16] one.

These strange attractors have a structure, and the structure is called fractal. Fractals are geometrical shapes that have a fractional dimensionality. One of the strong connections between chaos and fractals is that chaotic systems have strange attractors. In order to understand the order that lies within chaos we must understand the nature of fractal (or strange) attractors.

FRACTALS

A dimension is a way in which the size of something can be measured. For example, a box has three dimensions, which we call length, breadth, and height; or a human life has four, the three spatial dimensions plus time. The fewer the number of dimensions a thing has, the simpler it is both intuitively and descriptively.

We can generalize the idea of dimension to that of *dimensionality*, and say that any object has a dimensionality that is equal to the number of dimensions required to contain it completely. In the case of a Euclidean shape like a sphere or a circle, this is an integer, a whole number. For example, the circle requires two dimensions, so its dimensionality is two. Similarly, the dimensionality of a sphere is three, and that of a line, one.

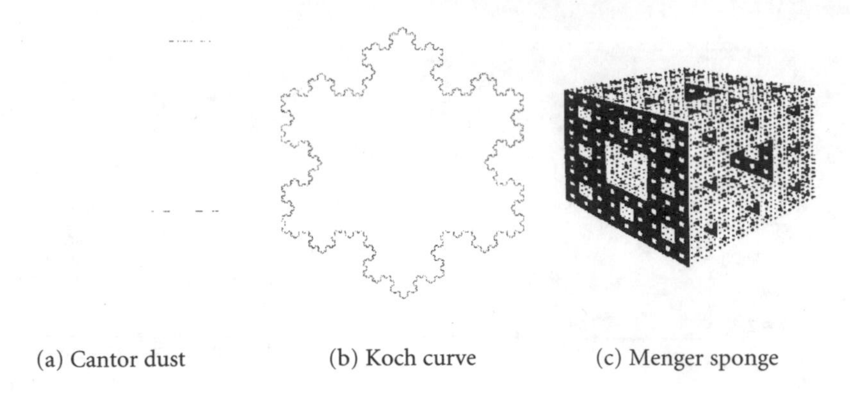

(a) Cantor dust (b) Koch curve (c) Menger sponge

FIGURE 4.6 Simple geometric fractals.

What is the importance of the relation of the number of dimensions to the dimensionality? For the Euclidean shapes the relation is 1:1 because we are dealing with whole numbers. When we turn to fractals, the meaning of the relation between the number of dimensions in which the object is contained and its dimensionality changes, because the dimensionality of a fractal is not a whole number, but a fraction.

It is difficult at first to know how to measure some of the properties of fractals. For Euclidean figures we measure length, breadth, or height, but for fractals that very first measure, length, is uncertain. An example of a naturally occurring fractal was given by Mandelbrot—the measurement of the coastline of a country like Britain—and is described in chapter 1. A mathematical fractal like the British coastline has an infinite length: its Euclidean length is not a way of measuring its dimensionality. In order to measure the dimensionality of a fractal we have to use a measurement that is independent of the scale we use. The dimension of a fractal is, appropriately enough, a fraction, not an integer.

In figure 4.6 you can see some simple geometrical fractals. All these fractals have their (fractional) dimensionality given below the diagram. The first, a very simple fractal called Cantor dust, is made by removing the middle one-third from a segment of a straight line. This operation is performed iteratively until (in theory) it has been done an infinite number of times. What is left is a sort of dust made up of an infinite number of very small line segments, which is fractal in distribution. The Koch curve, which we also met in chapter 1, is made by iteratively replacing a straight line segment with a four-segment line. By repeated replacements, a very complex curve of potentially infinite length can be constructed. The Menger sponge is made from a cube in which the middle ninth of each side is drilled away right through the solid. This is done repeatedly until the remaining "solid" is honeycombed with passages. Again, the internal area of the cube is increased by each operation until it too becomes infinite.

The way in which regular fractals are made is by replacing a section with a similar one at a reduced scale. Fractals like the Koch curve and the Menger sponge are made by means of replacement in this way. This is an iterative process, and iteration is the essence of fractal shapes. It leads to an important property of fractals: that of self-similarity. This means that fractals look the same whatever scale of enlargement is used to view them. This property is shared with real-world fractals, such as coastlines, and it implies something about the similarity, at different scales of operation, of the processes by which they are formed.

FRACTAL DIMENSION

Let us call the Euclidean space that just contains a fractal its *embedding space.* This space has an integral (whole) number of dimensions. For instance, the Koch curve can be wholly contained in two-dimensional space, and the Menger sponge in three-dimensional space. The definition that has been arrived at for fractals is that they have a dimensionality that describes how much of the embedding space they fill (this is known as the fractal dimension). This accords with the use of the term *dimension* in relation to Euclidean figures. Consider a circle: it covers all the points in its region of the plane and so has a dimensionality of two. A sphere covers all the points in a solid volume and so has a dimensionality of three, and so on. We can also construct mathematical objects such as hyperspheres, which fill four dimensions or even more. The dimensionality of such figures will always be integral, because however bizarre they may seem, these objects are still Euclidean shapes.

The dimensionality of a fractal is calculated from the number of parts replaced and the amount of scaling down.[17] This yields a fractional result, representing the fact that not all the points in the fractal's domain are occupied. At first the idea of fractional dimension may seem strange, but it is not so strange when you think about its physical meaning. Fractional dimensionality corresponds to the fact that a fractal is a thinned-out, tenuous thing in its embedding space. It has a density that is only a fraction of that of the available space. Cantor dust, for example, has a fractal dimension of 0.6309. This is less than unity, which fits the fact that the Cantor dust is less dense than a line (which has the dimensionality 1). The Menger sponge does not fill three-dimensional space; that is, it has less than three dimensions but obviously has more than two. The fractal dimension of the Menger sponge is 2.7268, and this seems in-

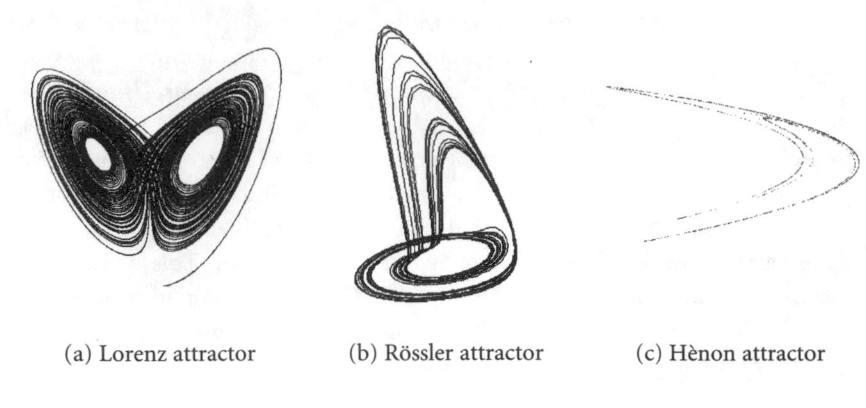

(a) Lorenz attractor (b) Rössler attractor (c) Hènon attractor

FIGURE 4.7 Complex fractals.

tuitively right. Real sponges are also fractals; that is, they are not space filling and have a huge (though not infinite) internal area. Similarly, the Koch curve is infinitely long, but bounded in a finite area, so we would expect its dimensionality to lie somewhere between one and two. It covers less area than a truly two-dimensional shape. The Koch curve's fractal dimension turns out to be 1.2618, and again this feels right. We can find some curves in nature, such as leaf boundaries, that have dimensionalities in this range.

But not all fractals are made out of straight lines, nor are all fractals regular. Indeed, regular fractals of the type we have been discussing are the exception, not the rule. Other kinds of iterative processes can produce fractals: many are created by chaotic systems. Some of these more complex fractals are shown in figure 4.7. The Lorenz attractor is a simplified model of the weather system discussed earlier. The Röessler attractor arises from another model of the weather. The Hénon attractor looks like planetary rings: it can be enlarged to show its self-similar structure. All these different models share one thing—they are chaotic and associated with them are strange attractors having a fractional dimension. This is a measure of how much of the embedding space is filled by the attractor and is called the *correlation dimension* of the system. The correlation dimension is one of the most important measures of the degree of chaos in a system, as we will see in chapter 5.

FRACTALS IN WONDERLAND

In the nineteen sixties a new mathematical object was plotted on a computer. The first drawings of it were crude and uninspiring and looked like uninter-

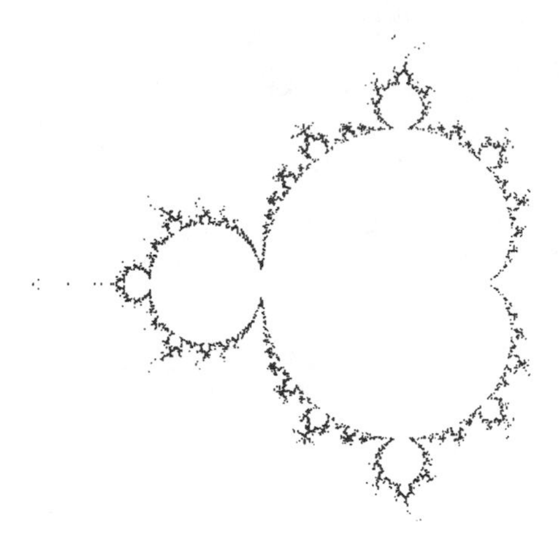

FIGURE 4.8 The Mandelbrot set.

esting blobs. Indeed, they were thought at first to be just the product of a computing error of some kind. When explored, however, this figure turned out to be a fractal of unlimited complexity. Iterated only 150 times, it looks like figure 4.8.

This figure came to be called the Mandelbrot set. One view of the Mandelbrot set is shown in figure 4.8. This is called an escape diagram and is the boundary between the points that stay inside the set after 150 iterations and those that fly off to infinity. The Mandelbrot set is certainly a fractal, and it shows a high degree of self-similarity when magnified.

The Mandelbrot set enjoyed a great vogue for several years. If the formula is iterated further, and if the resulting escape diagram is enlarged, the boundary takes on great complexity. If colored according to the iteration number when the points escape, this formula, one of the simplest possible in a plane, produces flamboyant and yet subtle shapes of extreme beauty.

Immediately, this shape looks of great interest to the biologist. Some people called it "the beetle" and it does indeed seem to have a head, thorax, and abdomen, as well as rather a lot of legs, or places where legs might grow. This resemblance between the Mandelbrot set and living things can be made even more explicit and we will look at how to do this in chapter 6. The Mandelbrot set is based on the simplest of iterative formulas and yet possesses potentially infinite complexity. Life is a process of growth, cell-division, and reproduction. Is it possible that in fractals the iterative nature of mathematics and the iterative nature of biology meet?

5

CHAOSTABILITY

Changing it rests. It is weariness to toil for and be ruled by the same.
(FRAGMENT 84)

CHAOS AND STABILITY

It is a paradox that chaos and stability, far from being opposed, are intimately linked together. In some ways they are even aspects of the same concept.

One of the most fundamental questions that we can ask about any system, complex or otherwise, is whether it is stable or unstable. For some systems this question is easier to answer than for others. If we are looking at a static system such as a pile of plates, it is easy to say that if the biggest are at the bottom and they get smaller as they go up, then the pile as a whole will be stable, and that if they are stacked in the reverse order, the pile is likely to fall over. But for other types of systems, such as dynamic systems, the question of stability may be much harder to determine. It is not easy to say, for example, whether a bicycle ridden at a particular speed and in a particular direction will be stable, and very often the only way to find out is to try it.[1]

Another problem is that stability for some systems is not easy to define. For organisms, the notion of stability is not nearly as simple as it is for a stack of plates or even for a bicycle. The term *stability*, when applied to living things, does not mean complete immobility: an organic system that is in stasis is dead. Stasis is indeed almost a definition of death. It is true that dead bodies decay, and that decay is change, but such decay is life at a lower level, that of the bacteria. The life of the proliferating bacteria begins once the life of the body is over; stasis represents the end of the line for any organic system.

It has long been believed that in a living organism the object of life is homeostasis (from the Greek: *homoio* = same). Homeostasis means keeping the organism's internal conditions stable and this is done by means of control mechanisms somewhere in the system. A good analogy with biological homeostasis is a thermostat: if the temperature of the room is too high, power to the boiler is turned off until the temperature falls; when the temperature becomes too low, the power is turned on again. It is worth noting that a thermostat does not maintain a fixed temperature, but allows change within certain boundaries. The temperature is not constant but it remains within acceptable limits.

Another critical aspect of homeostasis in living things is to prevent parameters from varying too widely to permit life. For example, if body temperature rises too high or falls too low, we will die. Our body temperature is not constant but fluctuates within an acceptable range. Variables such as body temperature (or more accurately, the amount of heat the body contains), blood pressure, and the levels of nutrients and hormones are all controlled in this way. It is quite normal for the temperature of the blood in humans and other homoiothermic[2] creatures to rise and fall each day. In humans the temperature may fluctuate by a degree or two, usually being lower in the morning and higher in the early evening. (Body temperature may also change for other reasons, such as hibernation and fever. Some of these changes have adaptive significance: fever is a way that the body attacks hostile viruses and bacteria.) These daily changes in temperature are part of what is called the circadian cycle, a rhythm with an approximately 24-hour cycle in which many biological variables change. This daily variation should not be thought of as an imperfection in the system. It is not a failure of the homeostatic mechanism that causes these things to change; on the contrary, the organism functions more effectively because they change. Homeostasis is not really what is being sought; it is *homeoscedasticity*, meaning "having the same variation." It is the range of variation that is held constant, rather than the variable itself.

As we saw in chapter 4, we can represent a cycle of change by plotting its path in phase space, and this is one way of showing the result of an iterative process. We can understand a lot about a system by examining the paths of its critical variables. The simplest path is a point; this represents stability, like that of the pile of plates. A more elaborate path is a limit cycle, such as a circle or an ellipse. An example is the planets, which follow nearly elliptical orbits around the sun. Still more complex trajectories are the result of chaotic processes and these may have fractal forms. We can reconstruct the behavior of a system from a series of values[3] representing outputs of the system, and Takens[4] has shown how it is possible to reconstruct the strange attractor of a chaotic system by means of this data alone. Using this method, the analysis of the characteristics of complex systems is often possible. Sometimes a phase portrait of the system attractor can be reconstructed; sometimes the parameters of the system can be determined; and in the case of some systems, it is possible to estimate their correlation dimension.

ATTRACTORS OF THE HEART

Chaotic systems have fractal (strange) attractors,[5] which have a fractional dimension. As we saw in chapter 4, we can measure the fractal dimension as an indication of how chaotic the system is. Since there is a strange attractor underlying a chaotic system, we can say that chaos also has a dimension, called the correlation dimension, which is that of the fractal attractor associated with the chaotic system.

The correlation dimension of a chaotic system is one of its most important features because it is related to how complex the system is and therefore how easy it is to forecast. Some systems change their dimensionality over time: if we examine living systems, we find a lot of variation.

One of the first indications of the effects of this variation was found by Ary Goldberger in his work on the heart. Goldberger and his colleagues[6] studied the heart and its rhythms as measured by the electrocardiogram (EKG), which records heart activity from electrical signals coming from the heart muscles. One of the measures that physiologists find most informative about the heart is the time between one beat and the next, known as the interbeat interval. This is never quite the same from one beat to another, but varies constantly. Goldberger found that the interbeat interval of the normal healthy heart is significantly more variable than that of the heart in certain conditions likely to cause heart failure. Majid[7] and his co-workers had already found that, in some cases of sudden cardiac arrest, the heart had a steady metronome-like beat; it was highly regular, while the normal healthy heart has a rhythm that contains much more variation. This is perhaps surprising at first; we have been led to think of an irregular heartbeat as an unhealthy sign. However, it appears that a very regular beat accompanies conditions such as congestive heart disease before death from heart failure. I will consider some possible reasons for this in just a moment.

Goldberger generalized this finding and hypothesized that the normal irregularities were signs of health in the heart function. Further analysis revealed that the normal changes in heart period were chaotic in nature and that there is a higher correlation dimension in these changes in the healthy heart than in the unhealthy heart. This led Goldberger to ask the question: is chaos good for you? This contradicted the conventional view, which was that regularity was the goal, leading to stability within the system. Goldberger's is a view that at first seems to be counterintuitive.

BENEFICIAL CHAOS

You might at first expect that chaotic variations in a biological system would be a bad thing, particularly in an organ as vital as the heart, and that the organism would try to minimize rather than attempt to maintain them. We usually think of things that are chaotic as being unpredictable and difficult to handle. How can this sort of variation be helpful to a living system? Surely what we want to do is to control, to stabilize, and to smooth out variation. But very often we hold to theories that are completely wrong: we need an intuitive leap to reverse them. A little more thought shows that our first intuition has gotten it completely wrong! On looking more closely we see that chaos is adaptive rather than the reverse.

One reason why chaos is a good thing in a biological system is that wear and tear should not occur in the same place but should ideally be distributed as evenly as possible. It is undesirable for the valves of the heart or associated arteries to be overstrained by repetitive stimulation, and an element of variation gives recovery time. Chaotic systems fit the bill very nicely: the wear is distributed in a much more even way than it is in a highly predictable system.

There is another more general consideration, however. It is good for living things to be chaotic because that is precisely what the environment is like, seen from the organism's point of view. The environment (which is not just one factor but a whole set of factors) bombards living things with stimuli, many of them threatening, some of them beneficial, in a pattern that is often hard to understand or even to perceive accurately. Seen in this way, it is clear that the organism must be capable of tracking this chaos by matching it with its own to keep one step ahead of possible threats, and to be able to model them perceptually.

Consider a bird flying: a seagull, say, soaring above the river estuary and staying aloft by steering into the wind. The air is in turbulent motion; eddies and currents cut across the main direction of laminar flow. To these cues, subtly felt and instantaneously responded to, the bird makes very slight adjustments to its wing angle, so that it cuts through the air—or the air parts itself against the wing—according as the sensed turbulence of the air pattern dictates. The correct cleavage plane in the air is instinctively and expertly found by the bird. The living system, the bird, is shaping itself to the chaotic patterns in the medium in which it flies.

This is not how an airplane flies, even though the wing angles of modern fighters are adjustable and the nose of the supersonic Concorde rises and dips during different phases of takeoff and level flight. Essentially, a plane forces the air out of the way and gives evidence of this by the demonic noise it makes in doing so.[8] Birds, especially sea birds, remain as efficient in high-speed flight as at low speeds, while planes at speed become much less efficient.[9] Drag is produced by friction against the surfaces that meet the medium, air or water. But if the right sort of turbulence can be induced, drag is greatly reduced. In a study by Sirovich and Karlsson[10] the wings and fuselage of an airplane were covered with small protrusions in an attempt to produce the correct kind of turbulence. If the protrusions were aligned in rows, the drag was actually increased. A random arrangement, however, reduced the drag. But what does "random" mean? In many cases it means that the plan of the arrangement was not known to those making it, so they cannot reveal it to us.

CHAOSTABILITY

The inputs to an organism from its environment are frequently chaotic, and for that reason are difficult both to perceive accurately and to respond to correctly. In the face of such a range of stimuli, what is the best strategy? To personify the situation, faced with an ever-changing world, we could try to maintain ourselves in exactly the same state—a sort of behavioral homeostasis. This means that our behavior would become rigid and stereotyped. But as we have seen, ultimately such rigidity is equivalent to death. If we maintain *too* steady an environment, then any variation outside of it will be more likely to lead to damage—rather like someone living in a steady warm temperature who catches pneumonia on first encountering the cold. Alternatively, we can allow ourselves to vary within a range of values, and this variation might be systematic or even chaotic. To maintain the maximum flexibility we should be ready to respond to whatever the environment throws at us: our degree of chaos should at least equal that of the environment. Stereotyped responses are highly maladaptive; flexible ones are likely to increase our prospects of survival.

Objectively, the organism, or indeed any adaptive system, is faced with the problem of survival. To achieve this it needs to be able to meet challenges from its surroundings. A system that is too rigid will not be adaptive enough to meet these challenges, but a system with an appropriate degree of chaos will be more flexible. Chaos in such a situation is a good thing. Without chaos we would not be living systems at all but mechanisms of a far simpler order, automatons that would have far less capacity for adjustment and adaptation than we have. Chaos is indeed a sign of life: it might almost be said that chaos *is* life; certainly it is very close to life.[11]

One form of adaptiveness is intelligent behavior. The nature of intelligence has been hotly disputed. Some people have even made statements like "intelligence is what intelligence tests measure," which fudge the issue. We no longer need to depend only on tests. It is now known that intelligence is also measurable by objective means: the speed of processing of perceptual information. If you show someone two lines for a very brief time and ask them which line is longer, then the shorter the exposure time needed before answering, the more intelligent the person is likely to be. Not only is there is a strong correlation between these inspection-time tests and conventional IQ tests, but inspection time may be a better measure of intelligence than IQ![12] Inspection time is strongly related to the time taken to form an internal representation of the scene; it may be easier for a more chaotic perceptual system to flip into the required representation. Perhaps the amount of chaos in the brain is related to intelligence too.

I want to suggest a more general form of Goldberger's hypothesis: that within certain limits, the higher the degree of chaos in a system, the more stable it will be. I will call this the hypothesis of *chaostability*. If valid, such a hypothesis has profound consequences for our view of life in two ways. First, it gives a view of living systems, that is, that they are high-dimensional chaotic systems. Second, it should dictate a philosophy of how to live life maximally, to enhance our chaotic dimensionality and hence our flexibility.

Supporting evidence for this hypothesis is not drawn simply from one study, but from many different types of investigation. When we examine various aspects of living things we find again and again that they display this same phenomenon. We even meet with it in systems that are not thought of as living but display the sort of adaptiveness we would normally associate with life. Such systems might, in a term borrowed from geology, be called *zoic*[13]—having lifelike dynamics and able to adapt themselves to life.

It is tempting to go further and identify life itself with intermediate levels of chaos. To do so would not be to explain very much, however; it would label and describe one of the essential characteristics of life, rather than giving the basis of an understanding of what life is. Since simply identifying chaos is not necessarily explaining how the chaos came about, its mechanisms are still unknown. It may be that there are universal conditions that govern all living systems, no matter how they are embodied. To see if this is so, we must look further into the nature of complex systems displaying chaos.

THE COMPLEX BRAIN

An electroencephalogram (EEG) is obtained by making electrical measurements on the scalp. The EEG is one of the most important sources of information about the activity of the brain and, not surprisingly, it has frequently been subjected to chaos analysis. The chaotic dimensionality of the EEG varies in different conditions, pathological and normal. For example, the EEG recordings of epileptic patients have been compared during a seizure and during a period of normal brain activity. The EEG has a dimensionality of about 4.5 in normal individuals, but this changes depending on the state, including the mental state, of the person. In epileptics the dimensionality is generally lower, around 4.[14] During a seizure it falls further, to about 3.5.[15] The finding is consistent with a simpler and less chaotic brain state in epilepsy as indicated by the lowered correlation dimension. Again, chaos is a sign of healthy functioning. Again, too, the conclusion is somewhat surprising, since people tend to think of epilepsy as chaotic and uncoordinated activity of the brain and limbs.

Walter Freeman[16] studied electrical signals from the olfactory bulb of rats. From the direct recordings he made he was able to devise differential equations that modeled the process of reacting to odors. When this model was evolved over time (that is, when the equations were iterated), chaotic attractors arose. According to this model, when nothing is presented to the sensory organ, the signals are characterized by moderately high dimensional chaos. When an olfactory stimulus is presented, the dimensionality of the signal falls and a recognizable attractor is generated, representing specific information about the odor and the context in which it was presented (e. g., whether it was accompanied by a reward or a punishment). The reduction of chaotic activity toward order was interpreted as the homing-in of the sensory system on the stimulus and it illustrates another way in which chaos can be adaptive: it acts like a "ready state" on which patterns (perhaps attractors) can be superimposed. This idea is reinforced by the work of Azouz and Gray,[17] who found that feature receptor cells in the visual cortex oscillate in synchronization with one another in a way that reflects the overall pattern being looked at. It suggests a hypothesis identifying perception with the presence of an attractor in the pattern of neural signals within the brain.

Shinbrott and colleagues[18] and other researchers have shown how it is possible to control a chaotic system by nudging it into a nearby stable orbit. Paradoxically, it is easier to control a chaotic system than a deterministic one, so a chaotic background is an adaptive one from the point of view of perceptual readiness. It is a classic finding in perceptual psychology that if you present pure auditory or visual noise , such as white noise or an evenly patterned visual field, to a subject, they begin to hallucinate quite quickly and will hear voices or see patterns or objects that aren't there. This is usually ascribed to the tendency of the perceptual system to project patterns on to the background noise, even when there is no pattern there, as though the brain needed to see patterns.[19] It seems that chaotic activity in a perceptual system similarly forms a noisy background against which, like the white noise background, a signal is more readily detected.[20] But a chaotic *imitation* of noise seems to be even more effective than noise itself, perhaps because chaotic systems can be easily displaced into a deterministic phase, and so lock on to a signal.

THE SIMPLEST SHAPE?

In view of the link between fractals and strange attractors, one might expect that fractals would be the natural objects of perception—it ought to be easier to see a fractal than a nonfractal. Remembering that the nervous system is itself fractal strengthens this conjecture. A fractal system should pick up fractal patterns more easily than nonfractal ones. It certainly does seem to be true that we have difficulty in seeing nonfractals. The arch nonfractal is the straight line: it is often thought of as being the simplest shape in nature. In fact, it is almost entirely missing in natural objects! If we exclude the edges of crystals, which are perceptually extremely short, there are no true straight lines in the natural environment. Straight lines figure largely in the class of optical visual illusions called geometrical visual illusions. In these we see straight lines as bent or as longer or shorter than they are or even as present when they are absent. Next to the line, the most poorly perceived figure is the circle, which is also the simplest fractal. Figure 5.1 shows a selection of these difficult-to-perceive figures. Is there something in the visual system of the brain that actually prevents us from seeing the straight line and its close relatives accurately? I shall return to this question in chapter 8.

If we think for a moment about how a straight line is made, far from it being the simplest shape, it is in fact one of the most complex. Straight lines are made by humans, and humans were made by three and a half billion years of evolution, so the straight line was the last shape to appear on earth.

It is not easy to make a straight line. We think it is easy because the equation $y = a \times x + b$, which is the formula for a straight-line graph, is a simple equation to write down. Although describing a straight line is easy, it is not easy to produce one: they are two different things. The simplest figure to produce by iteration is a circle; the next easiest is an ellipse. The straight line (unless you

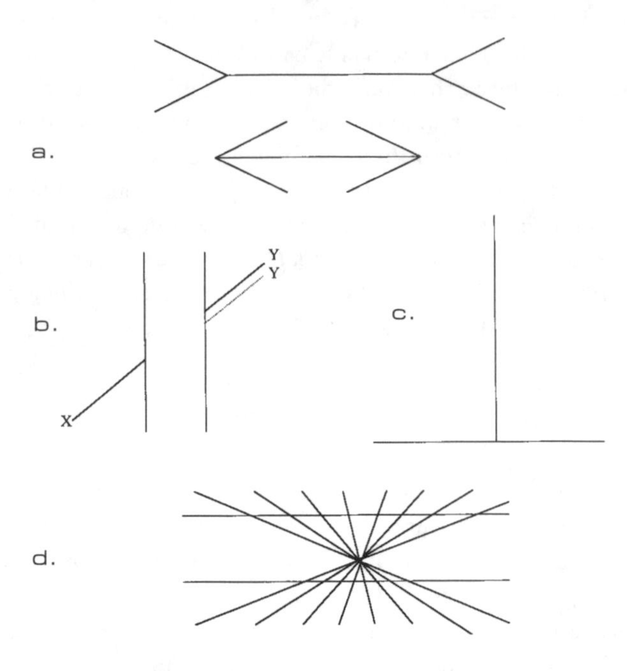

FIGURE 5.1 Some visual illusions. (a) Müller-Lyer illusion, (b) Poggendorf illusion, (c) horizontal–vertical illusion, (d) Hering illusion.

start with one) requires an infinite number of iterations. We could, of course, draw it with a ruler, but the ruler is already straight, so this is cheating—simply "photocopying" the line.

The evidence points to the fact that it is much easier to see fractals than straight lines; they are also more esthetically pleasing. The things we most enjoy seeing are natural objects: trees, rivers, fields and mountains, and, of course, people and animals. As Mandelbrot pointed out, nature is fractal: "Mountains are not cones, clouds are not spheres, . . . bark is not smooth, nor does lightning travel in a straight line."[21] These are the shapes that formed our environment during the evolution of humankind.

While some studies[22] have shown that there is no special perceptual preference for fractals in general as opposed to a random nonfractal, these "generic fractals" are not typical naturally occurring ones. Such mathematically produced fractals do not possess any recognizable form; they look much like scrambled eggs, and scrambled eggs have been stirred many times. The objects that make up nature have for the most part been "stirred" only a few times. These objects are some of the kinds of fractals to which our perception is especially sensitive and they are low-order iterates of fractal shapes, which will be discussed in the next chapter. Features of the human face are fractal, produced by low-order iterative processes. We are very good at perceiving other people's faces: we can remember hundreds or even thousands of them. Had the face been Euclidean, it is unlikely we would be able to remember so many, so easily!

CHAOTIC YOUTH; STABLE AGE

Another physiological advantage of chaos may be avoiding synchronization of firing in nerve cells. Garfinkel proposed[23] that periodicity in the central nervous system may be destructive, because if all the muscles were to contract simultaneously, spasm would result—Parkinsonism, for instance, may be such a case. One way of preventing periodicities from occurring or building up is to maintain the system in a chaotic state. Garfinkel also proposed that chaos was a feature more characteristic of youth, and stability of old age. This seems to be borne out by the data too, with EKGs of young people showing strong signs of greater chaos than those of older people.[24]

Glass[25,26] stimulated the heart cells of chicks at various frequencies, starting with a high frequency and gradually decreasing it. At high frequencies the heart beat once for every two stimuli; as the frequency was reduced, the heart beat only once for every stimulus; when it was lowered still further, the heart began to beat three times for every two stimuli applied. This is an example of what is called *phase-locking*, a phenomenon whereby two oscillators become coupled in a particular ratio. The same phenomenon occurs in mechanical systems too: a much-cited (but rarely seen) example is the shattering of a wine glass by the power of a female singing voice. A commoner example is the oscillation of the straps in a London underground train at certain critical speeds.[27]

Many patients have been fitted with artificial hearts, which work in parallel with their own.[28] By taking over most of the function of pumping blood, the artificial heart allows the patient's heart to rest and recover from damage. In one tragic case,[29] the patient's heart having recovered, the artificial heart was removed and the natural heart allowed to resume control. Within hours, arrhythmias had developed that could not be controlled and the patient subsequently died. The message of chaos in such cases would seem to be this: the system will adapt to the type of pumping it receives; if this is nonchaotic and regular, then it may be unable to make a rapid readaptation to a chaotic pump such as the heart is. In these cases some period of adjustment would seem to be necessary to allow a gradual transition from a nonchaotic to a chaotic regime.

THE MARKET WALKS

Economic systems such as markets and national economies are among the most complex systems studied, consisting as they do of vast numbers of interacting elements. Yet they show orderly features—a paradox that can perhaps be reconciled by their ability to produce order through self-organization. Markets may follow some of the same laws as fluid dynamics: analogies from fluid dynamics form a large part of the discourse of economic commentators, so the markets seem a fruitful proving ground for chaos. Expressions like *liquidity* to describe the ready availability of cash supply and *turbulence* to indicate rapid activity in the markets are the nearest analogies that can be found for the movement of money and they are highly suggestive of the kind of dynamics possibly at work. Whether particular markets show true attractors or not is a debatable question.[30]

The price of a share follows a quasi-random walk, but it has fractal properties.[31] There is also evidence of nonlinear dynamics during speculative market bubbles. This indicates that rather than being random, a share price might be part of a deterministic chaotic system and this leads one to the investigation of the parameters of the system.[32]

It is possible to study the dimensionality of share price movements either as a group, for example, an index such as the Financial Times or Dow Jones indices, or for a single share price. However, it is hard to find regularity in small share dealings (meaning a small company size or small dealings in a share, or both). In the case of small companies there are many internal and external factors at work—a change in management or the state of the market may alter the picture drastically. If trading volume is low, then even in a large company there may not be a flow of commodities sufficient to create a fluid-dynamic situation. But for larger companies where shares are frequently traded, such an exercise is possible, and it is also possible to treat currency and commodity trading in the same way. Despite the averaging of conditions across many shares that calculating an index entails, and that results in the loss of some of the source information, the analysis of indices like the Dow can also yield useful parameters.

A good test of changing degrees of chaos in the market under the impact of external events would be an extreme event such as a crash. At the time I was interested in the issue, one was readily available—the crash that occurred on "Black Monday" in October 1987. The object of the analysis was to see how dimensionality changed before, during, and after the crash. Four large companies and the Financial Times 100-Share Index were studied. In the epoch that included the crash of 1987, the correlation dimension of the share prices and the index fell from a characteristic 4.5 to 5 to a value of 3 and possibly lower during the critical period, rising again to 5 in the period after the crash. What does this imply? Chaologists are sometimes warned that it is meaningless to supply figures for, say, the dimensionality or entropy of a system in the absence of a model of the system: so, can these figures be interpreted?

When the analysis was first done I was surprised by the result, because I was expecting, in accordance with the traditional view, that chaos in market terms meant instability. Again, intuition has it wrong. The confusion results partly from the mental picture of the frantic activity of dealers in a falling market—the stereotypical idea of panic. But what is chaotic for the dealer or the shareholder may be quite orderly for the market. When the price is falling everyone knows it is falling; it is thus a deterministic situation, not a chaotic one. The problem is that everybody can't sell at once and this is what leads to their wild behavior. On the other hand, in normal times no one knows quite what to expect next, so the dealings go on at a moderate pace and the market price bobs up and down unpredictably in a pattern characteristic of "healthy" chaos.

THE CHAOS IN THE MACHINE

Can we take this idea even further? I have said that without chaos we would be not organisms but mechanisms. But what about mechanical systems themselves: are they really the models of determinism traditionally described, or does chaos play a role here too?

It appears that machines, particularly when they are running well, are not quite as regular in their action as has been thought. Here, too, we find that some degree of chaos is present in the more adaptive systems. I will give two examples of this. A study by Gonsalves, Neilson, and Barr[33] found that, if the modes of vibration of mechanical parts of a machine are chaotic, it is likely that this represents a more robust system than if they move in simple limit cycles, which would have a lower degree of chaos. The particular advantage seems to be in the likelihood of fatigue developing in metallic parts of the machine; the more chaos there is, the less the likelihood of fatigue developing.

My second example, although more anecdotal, is also from mechanics. In the last stage of generator manufacture, a "listener" used to approve each machine for shipping by listening to its rotor. Depending on the sound the generator made, the listener could say quite accurately whether or not it was likely to fail in use within a foreseeable period. The task of listening in one factory grew so much with increased production that more listeners were needed. The head listener was required to train them by means of the "sit-by-Nellie" method, where the trainee was told "that's a good one" or "that's a bad one." This was very time consuming, and it took about two weeks to train a listener by this method. After various attempts, a better solution was found: a neural network running on an ordinary PC was fed the generator noise from a microphone input. It was found that the network could learn in a few hours from the head listener's tuition and would subsequently do the testing reliably.

It is not altogether clear what was going on in this case, but it seems possible that it may have been the following. The sound of the machine must reflect, among other things, the way in which the rotor makes contact with its bearings. One kind of sound may be associated with many points of contact, another with only a few. The more points of contact, the more chaotic the orbit of the rotor is likely to be. It is common for the rotor mounting to be offset for optimum life of the bearings. This is an engineering rule of thumb having nothing to do with Newtonian mechanics, but the reason is probably to avoid setting up a limit cycle orbit on starting the motor. The more points of contact there are, the more the wear will be distributed among them. A point sustaining many impacts from the rotor will wear out more quickly, and this might produce rapid failure. It seems probable that it was, in effect, the degree of chaos of the system that the listener would hear. Might an experienced listener detect such differences? Given the lore, which is that an expert engineer can locate which particular bearing in a car engine is about to fail, it does not seem unreasonable to suppose so.

BREAKING THE BANK

The man who broke the bank at Monte Carlo was allegedly a certain Frederick Jagger, a spindle-maker for the cotton trade in Bradford in England. While visiting the casino he observed the turning of the various roulette wheels, and looking at a particular wheel, he said "There's something wrong with that spindle!" He placed his bets and won accordingly. It seems likely that what he saw was a reduced dimensionality of chaos in the wheel, the result of a faulty bearing, and from this he was able intuitively to deduce its probable orbit in phase space. This was essentially the basis of the technique used by Doane Farmer and his co-workers when in the early 1980s they attempted to break the bank at Las Vegas.[34] Using the new technology of microprocessors, they forecast the orbit of the ball on the wheel, assuming a throw of constant characteristics (if the croupier changes, you have to start again). These were not faulty wheels in the normal sense, but no wheel turns quite evenly, and therefore there is always a slight bias to be discovered.

It seems that mechanical systems are not so mechanistic after all! Might it not be that our traditional picture of an ideal machine is a rather poor model of what a good machine should be? In mechanical systems we repeatedly find that tolerances such as the offset of a rotor are built in, not because they are required by theory, but because they work. Lubrication, for example, is an entirely empirical requirement, designed to fill in the gaps between parts. Why shouldn't parts fit exactly and ideally-circular rotors move smoothly in their bearings? Because no engine could function in such a way. The circular path is not the ideal after all. In fact, what is happening is that there is a chaotic movement of the shaft against the bearing, making contact with points on the bearing in an irregular, ergodic[35] orbit, the impacts being further cushioned and distributed by the lubricant. No other system would suffice. In the end, any bearing wears out, but it does so sooner if the attractor is a simple limit cycle, the low-dimensional case apparently detected by the expert observer.

In the case of the heart, something similar may be happening, except that here the "sound" of the heart "engine" is the EKG record. A higher dimensionality in the EKG represents a heart that is using its components more flexibly. Just as in a machine, wear and tear should not always occur in the same place but should ideally be distributed as evenly as possible.

MUSIC AND MOVEMENT

Musical performance is another example of the tendency of adaptive systems to require a high degree of chaotic variation. It is arguably true that modern performances of classical music often lack a certain quality of variation that was present in earlier times, times closer to those in which the music was written. A comparison between the modern and older styles of playing can be shown in a study of recordings. If you listen to the followers of the Polish violinist Auer, you will hear a quality that is markedly lacking in the style of the more modern instrumentalists. Comparisons may also be made between living musicians of different traditions. Oscar Shumsky, for example, is an example of a modern violinist who learned from Auer himself. Shumsky expresses the same quality in his playing that is heard on some of the early recordings. The Polish-born pianist Horszowski died recently at the age of one hundred and one, following a period in his nineties in which critics thought that he did his best work. Horszowski learned from a pupil of Czerny, who was in turn a pupil of Beethoven. In a tribute broadcast a few months after Horszowski's death, the pianist Radu Lupu likened the quality of his rhythmical interpretation to the beating of the heart. What is this quality, which the older musicians had, and which is essential in some way to bring out the full feeling of the music?

The most striking feature of this kind of playing is its variation, whether of tempo, loudness, or pitch. First, consider tempo. The subtle variations in tempo in early recordings add to the life of the music in a way that has been lost by later instrumentalists. These changes in speed, once thought so essential, are now more often thought sloppy and romantic. Variations in tempo are referred to as *rubato,* an Italian word meaning "I steal." The idea behind applying this word to changes in rhythm and tempo is that what is taken away from one bar or interval between notes is paid back in another, while the overall sense of rhythm is not lost. The robbery is said to be an "honest robbery" (to quote Scholes' *Dictionary of Music*[36]), although the matter appears to be not quite so straightforward, as often the changes are not repaid or compensated for within the bar. There was quite a debate about rubato in the nineteenth century and Berlioz declared that in the end it was "taste" that should be the arbiter, in other words, a perceptual appreciation of what life-like music ought to be.

This sort of playing lends an elasticity to the rendering of the piece that gives it a life of a kind that metronome-like regularity lacks. The result is added life in the performance. Nowadays a very different style generally prevails. The aim of many musicians is apparently to give a fast, accurate rendition of the score as it is printed; the flautist James Galway has said of his rise to success that his aim was to play more notes to the minute than the opposition. What is lost in this approach? Rhythmical variation is an early casualty; exactness is stressed at the expense of expressive quality. In general, rates of note-playing seem to have climbed in recent years and the execution of difficult passages is successfully achieved with greater apparent ease. But is it notes per minute that is the criterion of excellence in performance, or is it not the spirit of the music—how alive it is and in what way? Many modern performances lacking rubato are effectively dead and we lose the opportunities that life offers when we listen to them only.

Another factor present in these older violin performances and lacking in contemporary versions is *portamento,* which means carrying the pitch from one note to another. When an interval is played, a long scooping effect is sometimes obtained, which is unfashionable now. Yet with some listening practice, this sound becomes familiar and seems to add something to the performance. Many contemporary violinists play in a much more staccato manner. Again, listening and making the contrast between the two styles gives the feeling that something has been lost.

Many psychologists take the view that any perceived emotion is the product of two components: arousal, the experience of raw undifferentiated feelings;

and attribution, perceiving the "cause" or source to which we attribute the arousal. The cause to which we attribute the emotion does not have to be the cause of the arousal. If the room is hot, we may feel irritable but attribute it to some extraneous cause. It is no accident that so many arguments, as well as so much merry laughter, may break out over meals. Eating food arouses us and the arousal produced may be attributed positively or negatively to other people's behavior. These attributions may frequently be made as a result of listening to music such as background music and pop music, especially if these are heard out of context. Just as the effects of undue regularity are apparent in the interpretation of classical music, so they are in popular music. These effects often consist of stimuli having induced or artificial regularity. A beat produced by a synthesizer is usually absolutely regular, and this sort rhythm of induces marked physiological effects. Increased arousal, raised blood pressure, and increased adrenaline levels are typical of someone listening to music of a very regular kind. That is why strict tempo is preferable for dancing, where the extra energy is needed. But in contemplative listening or as background music for conversation these effects are undesirable; a raised level of physiological arousal will not usually be attributed to the music but to external causes. The potential for tension and irritation is enormous.

Plato prescribed that in his Republic certain musical scales would be reserved for special occasions and some even banned altogether as socially undesirable. Plato's philosophy has sometimes been condemned in modern times as fascist, but perhaps he had some good ideas about social engineering too.

DIGITIZED SOUNDS

Methods of musical reproduction have also changed and this has had an effect on another quality of musical sound—pitch. The most frequently used medium now for listening to recorded music is the compact disk. This is a digital technology in which the sound is converted to a series of binary signals (on or off bits) that are cut into the surface of the master disk by means of a laser beam. In this way, a musical sound, which is continuous, is turned into a series of quantized pulses to the ear that it is hoped will approximate the original sound wave. The range of pitches reproduced is very large—well beyond the upper threshold of most people's pitch perception—and it might be expected that this could introduce no perceptible effects. However, because of interactions between the inaudible frequencies and the audible ones, musicians would easily detect the results of this digitization process in the sharpness of the upper registers, so a frequency cutoff has to be imposed, effectively blocking the high frequencies. For this reason, many hi-fi enthusiasts still prefer vinyl disks, which use analog reproduction methods.

The same kinds of effects are entering the visual media as are found in music. Much of what people see today is on television or computer screens, and although the cinema is now in a state of relative recovery, its decline over the past few decades has been considerable. Both computers and televisions use technology that quantizes the display into dots; the same is true of screened printed pictures in newspapers. This means that instead of seeing an artist's impression, which is capable of showing continuous gradients of light and shade and hue, we are seeing a version with a very much reduced number of elements, and those elements have sharp edges. This produces two effects: the scene is divided into fewer and more regular elements than would be the case in a drawing or painting; and the sharp edges artificially induce high spatial frequencies[37] that are not there in the original. In addition to this, the picture flickers on and off regularly every sixtieth of a second. The results of this kind of very regular photic stimulation are that people are at once visually fatigued and visually deprived. In extreme cases this condition becomes pathological, for instance, when flicker induces an epileptic seizure.

Analog reproduction is true to the original in ways that digital reproduction at present levels of accuracy cannot be. Sympathetic interpretation is undermined by competitive musicianship. The effect of all this is that there are powerful forces working against lifelike, or zoic, variation in the artistic media.

THE GAIA HYPOTHESIS

Some of the most puzzling questions facing humankind today concern the stability of our environment on earth. On the one hand we can see that human activities have despoiled huge areas of the earth's surface. Direct interventions such as large-scale deforestation, the depletion of marine food stocks, the effects of human activity in spreading deserts, the production of pollutants, and the emission of vast quantities of carbon dioxide, are by now familiar examples from the litany of environmental interference that has been wrought by humankind. On the other hand, despite all of this, the earth withstands these attacks. Such indeed is the resilience of the earth to the insults heaped upon it that some scientists have postulated feedback mechanisms that are capable of preserving a balance, even against what are apparently the best efforts of human beings to upset it. One scientist, James Lovelock, has suggested the bold idea that the whole planet forms a system that can be regarded as a single organism—that the earth is alive. This is known as the Gaia hypothesis, named after the ancient Greek goddess of the earth. Seen in this way, what appears to be happening is a struggle for survival by the earth against a terrible assault by her offspring.

What exactly does the Gaia hypothesis entail? If the planet as a whole is a living organism, then its first concern must be to maintain the conditions of its own life. In the case of the earth, this means that the shell of the ecosphere surrounding the planet is of primary concern. The rest of the planet is not irrelevant or unnecessary to the maintenance of life. Conditions must exist for the retention of an atmosphere, for adequate geological structure, and for warmth; and this implies a globe that is of sufficient mass, but not too large, and geologically active, but not too active. These conditions could be likened to the skeletal structure of the organism and would be as necessary as a skeleton is to a vertebrate. However, it is the skin, the ecosphere, the layer of warmed and moistened earth inhabited by the living creatures of the various kingdoms, that must be the primary focus of interest from the point of view of the Gaia hypothesis.

To say that the ecosphere is alive must mean that there is a super-organism composed of all these separate organisms, just as the Portuguese man-of-war jellyfish is a super-organism or human beings are super-organisms. All the parts of a living thing do not need to be connected by continuous chains of matter. If that were so, our nervous system would not be the functional unity that it is. A jellyfish can pass through a relatively fine sieve and remain intact. What defines an organism is its functioning as a living thing. If all the earth's living things are to form a super-organism, they must act as cells in a body do, communicating and interacting in a sustainable way. This may be the function of the food chains and symbiotic connections, which exist plentifully in the natural world. It is then a question of whether this symbiosis constitutes the totality of biota or least enough of them to make a super-organism feasible and whether the manner in which the total system interacts is self-maintaining.

There is ample evidence that these chains exist and extend throughout the natural world. Many species coexist with or prey upon other species, living in dynamic equilibrium with them. The evidence that this whole system constitutes life is equivocal, since the conventional biological definition of life, which includes the criterion of reproduction, obviously does not apply. The earth does not reproduce itself; if it is an organism, it is a single, immensely long-lived one. In the definition of super-life other criteria must be critical, such as the existence of stabilizing mechanisms that will correct any long-term imbalance and prevent the system from going off the rails–the "death of earth." Arguments have gone on about the role of plankton, the regulation of the planetary carbon budget, and the necessity for large areas of rain forest, but the debate has been largely inconclusive.

The ecosystem does indeed seem to be well knit, but whether the conditions are sufficient to fulfill the criterion of a single living organism has not been settled. It is still not known, for example, whether the oceans are a net carbon source or a net carbon sink. It must be remembered, however, as a counter to the rather dramatic view of the death of earth scenario, that while events such as extinctions of whole groups of species—extinctions that may be a regular feature of life on earth—are catastrophic happenings from the point of view of the species themselves, they are not necessarily so from the point of view of a super-organism. The individual cells are not what count; it is the overall level of organization and its maintenance that is the essential defining feature of life. Whole classes of cells can become vestigial without the organism suffering any loss of viability, and similarly, even mass extinctions need not and apparently have not threatened Gaia.

CHANGING CLIMATES

However, whether or not the earth is an organism, it certainly appears to be both resilient and stable. Mother earth, despite the assaults upon her, maintains a kindly environment for humankind. Indeed, it is remarkable that climatically we appear to be living in an exceptionally well-favored period. Since the last ice age the climate has been unusually stable, especially during the last eight to ten thousand years. We know this because the record of the past weather is preserved in the ice that has built up in polar regions. In Greenland and other places deep borings have been made into the ice and very long ice cores extracted. The Greenland Ice Core Project (GRIP) core contains many layers of varying thickness and constitution, which are the historical record of annual ice deposits. The length of these cores represents access to periods correspondingly far back in time.

Recent borings[38] have produced layers that were deposited up to a quarter of a million years ago, although the layers become more compressed as they go further down, making them harder to read. When ice is formed, air is trapped in it. The amount of oxygen isotopes in the air in each layer can be related to the probable air temperature at that time. By this means it is possible to reconstruct past weather conditions, at least as far as annual temperature goes. Very interesting findings have emerged from this work.

Until about ten thousand years ago the climate typically underwent wild swings of temperature, with ice ages or glacial periods and relatively warm spells (called interglacials) between them as the extremes. For long periods the earth was very cold indeed. The interglacials are usually much shorter than the glacial periods themselves. The last such glacial period lasted for about one hundred thousand years and ended only twelve thousand years ago. On this basis one might expect that we are due for another ice age at any time, which would mean a blanket of ice over much of the habitable earth and, if it were anything like the last one, extending in the Northern Hemisphere down from the pole to cover much of Britain and North America. But the situation is more complex than simply a variation of temperatures within a certain range; there are strong patterns within these periods too. What has emerged, not only from ice core data, but from other proxies for climate, such as lake sediments, is that the last interglacial period was not like the interglacial in which we are living now, which is known as the Holocene. During the last interglacial, called the Eemian, the climate was typically far from stable, with large variations in temperature, of a much larger order of magnitude than we experience now. For example, about 130,000 B.P., average global temperature changes of the order of ten times the present ranges were common. Even after the end of the last ice age, there were sudden swings back into glacial conditions lasting for hundreds or thousands of years.[39]

What kinds of temperature changes are we talking about? Nowadays the variation about the mean position is a degree or two in a century. (If this does not sound like much, remember that it is a global average. If the average global temperature changes by only one degree, then local temperatures will vary much more.) Even such a small change is large enough to produce significant hot or cold spells, like the summer of 1994 or the winter of 1972. At extremes, events like those of 1816–the so-called year without a summer, when snow fell in New England in June–may occur. You have only to imagine the effects of an unusually cold spell in terms of burst water pipes and power failures, or those of an exceptionally dry period on agriculture, to see the outcome of such "small" variations in climate. Moderate variations would render communication and transport all but impossible. But these extremes are not even remotely comparable to what would result from a global change of 10°C. This sort of variation in either direction would wipe out human populations in many areas and make all but the most primitive way of life impossible for others. It would incidentally cause mass extinctions of many other species on which civilization presently depends. Such variations were, it now appears, a characteristic of the last interglacial period. Human life at such times must have been a tenuous business.

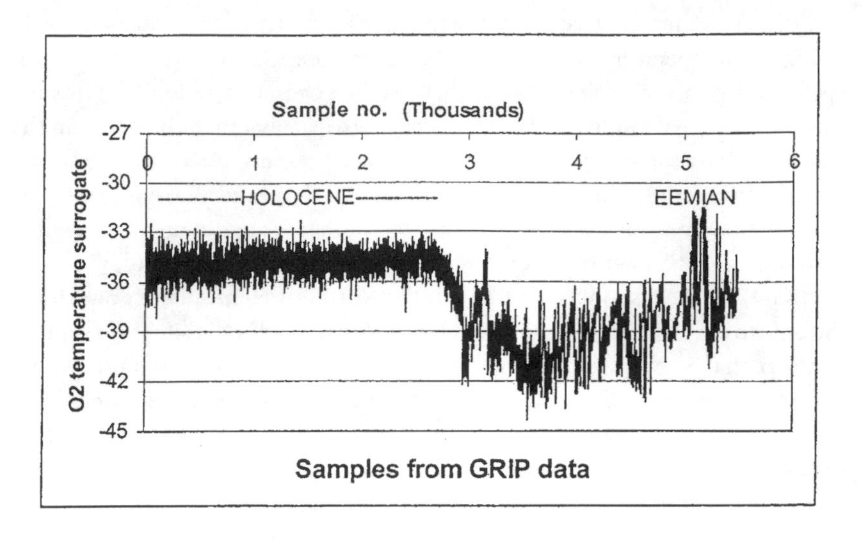

FIGURE 5.2 Greenland Ice Core Project data, showing oxygen isotope variation over the last 250,000 years.

OUR GOLDEN AGE

What is truly remarkable is that all this has changed and the climate is now in a singularly tranquil and uneventful phase. Some data from GRIP are shown in figure 5.2, illustrating that the Holocene experienced remarkably little variation. During the last ten thousand years global temperatures have changed by only about the levels we have experienced the twentieth century—a fraction of a degree—an amazing record of stability.

If the climate is still one of the most important factors in our lives, since it is essential that there be a moderate range of temperatures, sufficient rainfall, and so on, it must have been even more critical at the beginnings of civilization, particularly in the early, emergent stages. Without the right weather conditions, systematic agriculture is impossible; for example, it must be possible to preserve seed-corn for the following planting season rather than eating it, if there is to be any surplus food. Also it would be impossible to have a marketplace for the exchange of goods and services if the climate did not allow regular communication and travel. In these times it is to some extent possible to defy the elements by human ingenuity and construction. But events like the moderate hurricane that struck the United Kingdom in 1987 raise doubts about the ability of our civilization to survive for long in conditions even slightly different from what we regard as normal. Nowadays, as in prehistoric times, any departure from a fairly regular cycle of weather patterns would be potentially ruinous. In other words, civilization would never have developed at all without the stable weather conditions we are now experiencing. Moreover, those conditions took only a short time to be established. The change from the previous regime to the present one came about in a very short period–something like forty years.[40]

In view of these considerations, it has been suggested that our civilization is resting on a knife edge. It has been possible to reach the level we have only as a result of the relatively stable climatic conditions of the last ten thousand years; a return to the previous regime would obliterate our civilization altogether. It is argued that any environmental tampering, such as the release of extra quantities of carbon dioxide, might prove to be a destabilizing influence, so it would be best to leave well enough alone. This is the "Do not adjust the controls which have been pre-set in the factory" school of thought.

In the light of the previous discussion, another alternative is at least possible, and with our current ecological concerns in mind it is worth very serious consideration. This is that we may once again have gotten it completely backward. Rather than conditions of human life having stabilized because of the climate, the climate may have stabilized at least partly as a result of human life. It may be that human beings have helped bring about the conditions that now favor their existence. It is only in the last ten thousand years or so that we find evidence of human interference with the environment in ways that might be significant. These include the use of fire to destroy forests and in slash-and-burn agriculture (in which the grain is harvested with a stone sickle and the scrub is then set alight, to provide carbon and other organic deposits for later fertilization of the area). The herding of cattle would create zones of high methane emission. Such activities alter the climate in significant ways. The reflectance of the ground is a significant factor in global heat uptake and precipitation, and the altering of the atmosphere by the addition of gases like carbon dioxide and methane is another factor in weather. If it is tempting to reject such an idea, perhaps it should be remembered that until very recently, no one seriously supposed that humans could by their activities influence the climate. Now anthropogenic effects on the climate are established beyond any doubt. Is it not worth considering how far back in time this influence might have extended?

MAN THE CURATOR?

It is at this point that the role of chaotic systems becomes relevant. To say that the present climate is uneventful is not to say that it is predictable. Indeed, we know the difficulties of short term-weather prediction, on which so much effort is spent. Before 10,000 B.P. the situation was different in this respect, too: the weather then was likely much more predictable. Analysis of data from the GRIP project indicates that the climate in northerly latitudes between two hundred and fifty thousand and ten thousand years ago was significantly less chaotic than it has been since ten thousand years ago.[41]

Does the recently stabilized climate correspond to a state of moderate-dimensional chaos? It has been proposed by Nicolis[42] that there is a climatic attractor, and although this has been contested, there are certainly short-term attractors in the weather, for example, governing wind speed.[43] A problem is that if the climate is indeed a chaotic system, one might expect to see a degree of self-similarity in its variation over both the long and the short time scales. Unfortunately, the correlation dimension of the long-term attractor is low, about 3, while that of the short-term attractor is much higher, more than 7. One possibility is that the dimensionality is different now from what it once was. We may be living in a period of remarkable calm, not because of a lack of chaos, but because there is an enhanced level of chaos in the climate of recent times, say the last ten thousand years. This fits, since during the ice ages there were wild swings of temperature up and down, while modern times have been characterized by many small changes of the type we associate with higher-dimensional chaos.

This may begin to sound familiar: once again the conclusion is the same—that a chaotic system is healthier, from its own point of view, than a more deterministic one. As we have seen, there is a strong association between a moderately high-dimensional chaotic system and one that is life-like–a zoic system. This may also apply to the system—the huge biosystem—of the earth. A reasonable diversity of human activity in interfering with "natural" resources is to be desired and maintained. The production of anthropogenic carbon dioxide, for example, may be a necessary part of the stabilization of the planetary climate, essential to prevent the occurrence of another ice age, followed by another interglacial with its traditional extreme swings of temperature. The predominance of huge uniform areas of land and sea is a state that will lend itself more easily to the large steady swings and movements that represent, from humankind's point of view, a series of catastrophes. The division of the ecosphere into many local regions, diverse in their activities, is likely to set up the kind of chaotic system characterized by many small changes that preserve stability.

From this viewpoint then, the earth is indeed now a living system, or much more nearly so, due partly to the activities of mankind during the few hundred years around 10,000 B.P., which helped to stabilize the weather. Far from being the destroyer of the environment, humans may be its creator, or at least its curator. And far from being potentially destructive, human activities may be essential to preserve the stability that the recently born organism Gaia needs. If this is anywhere near the truth, then we must indeed be very careful about upsetting the balance of things that has been achieved, but not in quite the way that has been promoted by the traditional ecological movements. If, far from destroying the environment, human activities such as industrial activity, agriculture, and even desertification have helped to stabilize it, then the prescription for ecological health would be very different from the courses of action advocated by environmentalists. It might be very necessary to maintain diversity and activity, including the kinds often condemned—industrial emissions, deforestation, desertification—provided they are kept at the right level and are reasonably distributed. What is unwise in the extreme is to behave as though a question has been solved when, in fact, it has not been solved at all.

LOST OPPORTUNITIES?

It is fascinating to discover that there was another period of relative stability in the remote past, but within the probable existence of modern man, and now brought within the compass of the ice records. That stable period was around one hundred thousand years ago at the end of the Eemian, and it lasted for about two thousand years, within the possible period of the existence of modern man as defined by the paleontological and genetic evidence. Could it be that there was an attempt before our own to start a civilization—an attempt that foundered disastrously?

Many scenarios are possible. It is possible that other protohumans, such as Neanderthal man, who coexisted with modern man for some thousands of years, were forerunners of modern man in this respect also. Was there a Neanderthaler civilization that was the precursor of ours? What level of civilization might have been reached in this period? What ended it and returned the earth to an ice age that lasted until very recently? Was it an uncontrollable change emanating from purely within the system of the earth? Or could the earth have been knocked out of balance from without, say by a cometary collision—an event we now know to be fairly common? Such an event might well produce a huge dust-cloud "winter" lasting for many years—long enough to very probably de-chaostabilize the climate. Or was the return to glacial conditions brought about by some sort of self-terminating hominid behavior? Perhaps we shall one day have the evidence to decide these questions.

CONCLUSIONS

I now wish to draw the discussion of these areas together and state a general hypothesis. I propose that adaptive systems are those capable of attaining a degree of chaotic stability. Chaos in a system can be seen as a way of spreading its loads and strains more evenly. Chaos in this view is characteristic of adaptive systems, where adaptiveness may be independently identified by the criterion of the survival of the system.

There are three ways in which chaos is adaptive. First, a chaotic system is less easy to forecast than a nonchaotic one, and in the case of living organisms this is obviously an advantage. If your enemy can guess what your next move will be, it is easy for him to defeat you, whether the game is one of chess or life and death. We say of people that they have lost their temper—this sums up very well the lack of flexibility that enters the behavior of someone who is angry and to that extent out of their own control and within that of a calm opponent. Second, chaotic systems are more stable: they have less long-term variability and more short-term variability. They are less liable to the wild and sudden swings that characterize simpler systems. Third, a chaotic system causes less wear and tear to itself: it is more viable. The heart distributes strain evenly among its valves and the circulatory system reduces the pressure of pumped blood, which would otherwise rupture the blood vessels, by its branching structure. Fourth, it is easier to control a chaotic system with a minimum expenditure of energy. We could sum these advantages up in the concept of chaostability. The traditional view of living things is that they aim at stability through homeostasis (constant internal conditions): for this concept we should perhaps now substitute chaostability in a context of constant internal variation.

The hypothesis of chaostability has profound consequences for our view of life in two ways. First it is a view of living systems; that is, that they are high dimensional chaotic systems. Such systems are more adaptive and can respond well to exogenous shocks to the system. Second, it should dictate a philosophy of how to live life maximally: to maintain our chaotic dimensionality and hence our flexibility.

Heraclitus said "They would not know the name of Dike if these things did not exist." The word *dike* according to Kirk means something like a rule or measure. It is probably cognate with the Sanskrit word *dikr*, meaning the rhythm of the heart. In Arabic, a language with a rich variety of word meanings, *dhikr*, according to Baldock[44] means the remembrance of God through repetition of the divine name. According to Ibn 'Ata'Illa, "The dhikr of the heart is like the humming of bees, neither loud nor disturbing."[45] We know in our hearts what is dynamically true.

6

THE GEOMETRY OF LIFE

The real constitution of things is accustomed to hide itself.
(FRAGMENT 123)

FORM AND FORMULA

Life is a complex phenomenon—the most complex known in the world—and in order to explain the forms of life we must find a process that is either equally complex, or is simple, yet capable of producing great complexity. For the last 150 years it has been the accepted view that the chance patterning brought about by the forces of natural selection, mutation, or some combination of the two are such a process, but as we saw in chapter 2, there are several problems with this view (the neo-Darwinian synthesis). It is now becoming clear that the complex and well-defined forms of biological organisms cannot be accounted for on any realistic time scale if the processes of natural selection are considered alone and without any directing influence.

These difficulties are made worse by the direction of current thinking in biology, where the chemical approach has yielded plenty of correlations between genotype and phenotype but no functional relationships, because the intervening steps are merely causal. At the same time, science, and especially life science, is rapidly coming to recognize that it is dynamic processes, rather than causal chains of events, that drive the world. Is there a way in which these viewpoints can be brought together so that the science of biology can progress? I believe they can, but such a synthesis implies no less than a paradigm shift in our thinking, the consequence of abandoning a random-selection worldview in favor of an iterative-sequential worldview.

In this chapter I will try to set out a theory of the forms of life that would account for the facts better than neo-Darwinism alone. The theory I want to

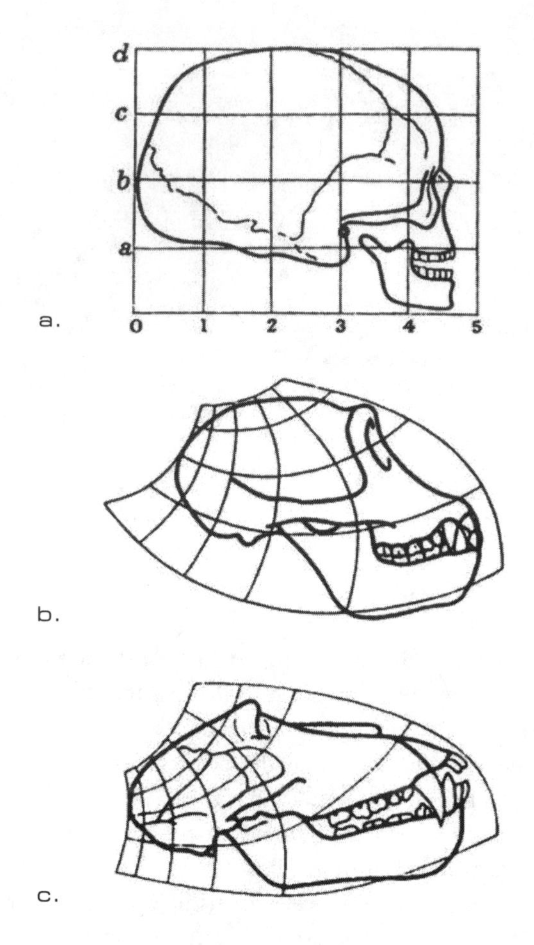

FIGURE 6.1 The skulls of (a) a human, (b) a chimpanzee, and (c) a baboon compared in different coordinate systems. [Reprinted from D'Arcy W. Thompson, *On Growth and Form* (Cambridge England: Cambridge University Press, 1942), with permission of Cambridge University Press.]

suggest is that the shape of living things is the result of iterative processes taking place inside each cell of the organism. Many of these processes are nonlinear, which implies that they may also be chaotic, although there must be linear processes involved as well. It is these nonlinear processes that determine cell division and thus the shape of the organism and its constituent organs. There are many pieces of evidence for this suggestion. None of them is by itself conclusive, but taken together they make a picture of overwhelming likelihood. I will review some of these, concluding with a description of some models of nonlinear processes, models that are at present rather primitive but point the way toward what might eventually become a much more complete understanding of morphology.

There is certainly plenty of evidence for mathematical processes at work in biology. The phenomenon of invariances—features of living things that are virtually identical from one representative of a species to another and have remarkably constant mathematical properties—for example, the angles at which branches leave the trunks of trees and the angles between interlocking bones of the human body suggest a mathematical rather than chance origin of their form. That there are manifold regularities of form in nature was recognized half a century ago by D'Arcy Wentworth Thompson in his monumental work *On Growth and Form.*[1] In the closing chapters he sets out a portfolio of biological forms that he could demonstrate were mathematically related to one another by simple transformations like stretching or bending the coordinates of the frame on which they are drawn. Seen in this way, the skulls of a baboon, a chimpanzee, and a human are the same when they are drawn on different coordinates (see figure 6.1).

Thompson believed that, rather than living things having accidental shapes, there was a mathematics of morphology, an idea that is just beginning to be explored. John O'Connor and Edmund Robertson at the University of St. Andrews have written software[2] in which many of Thompson's original results have been reproduced using quadratic mappings, a technique different from his but yielding similar results. The user enters parameters that will transform pictures on a computer screen, thus transforming morphologies at will.

FRACTAL PATTERN IN NATURE

If there is a mathematics of living things, then the most likely candidate is the mathematics of fractals. Why this is so becomes clear when we consider how fractals are created. Regular fractal patterns are made by repeatedly substituting scaled down subsections of the original pattern. As we saw in chapter 1,

FIGURE 6.2 Minaret cauliflower at three successive magnifications (a–c). (Reprinted from F. Grey and J. K. Kjems, "Aggregates Broccoli and Cauliflower," *Physica D* 38 (1989): 154–159, with permission.)

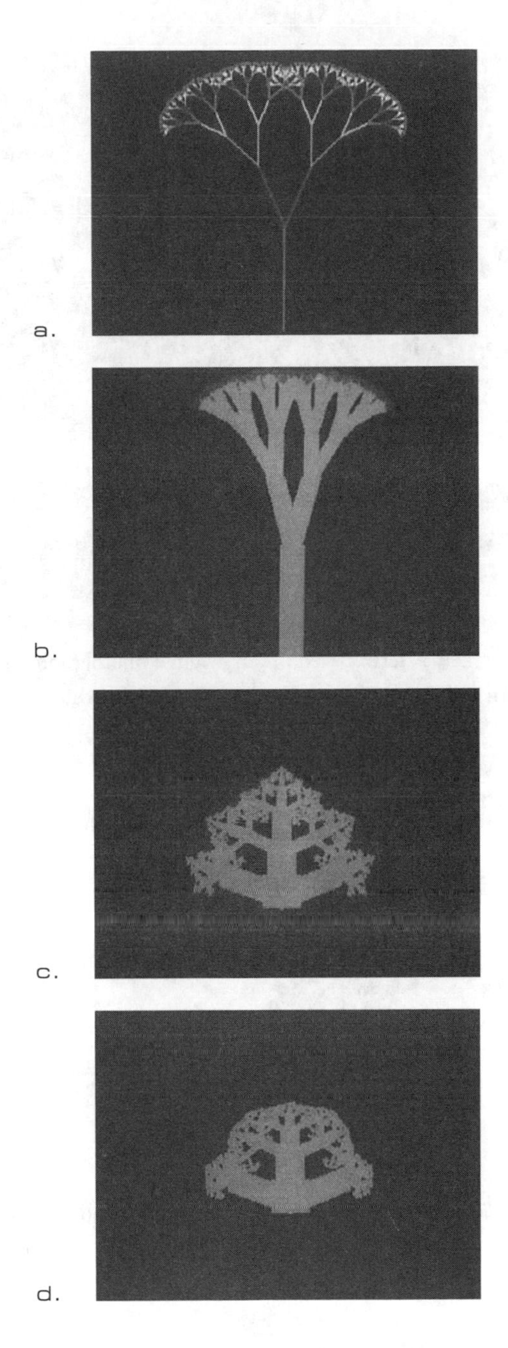

FIGURE 6.3 Models of hemlock (a), broccoli (b), minaret cauliflower (c), and hemlock (d). [Reprinted from F. Grey and J. K. Kjems, "Aggregates Broccoli and Cauliflower," *Physica D* 38 (1989): 154–159, with permission.]

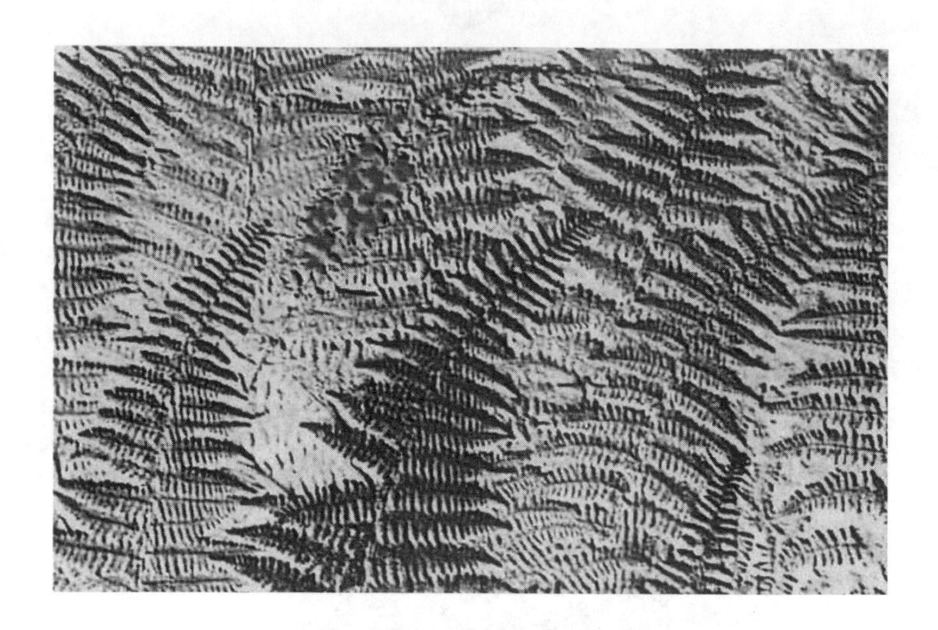

FIGURE 6.4 A fern. Based on an original photograph, "Fern and Purple Loosestrife," by David Fingerhut.

FIGURE 6.5 Computer simulation of a fern pattern.

what happens is that a chosen section of the pattern is replaced by a new element and this replacement is repeated at each successive level, producing the phenomenon of self-similarity. This process can be continued to infinity (in theory at least) or it can be terminated after a certain number of iterations.

Many examples of self-similarity can be found in nature, especially in the shapes of plants. Cauliflowers of the type known as minaret display self-similarity to a remarkable degree of detail. Figure 6.2 [3] shows a minaret cauliflower at three different magnifications; it looks roughly the same at all of them. This cauliflower is fractal to about 3 orders of magnitude or 6 branchings. Other related species share this characteristic of self-similar repetition of form. Computer models of the underlying fractal structure of the minaret cauliflower plus that of broccoli and of a tree trunk are shown in figure 6.3. All three have a branching structure but with different numbers of branches and different thickness of stem.

Fern leaves resemble their individual constituent fronds, and the fronds in turn are like the leaflets of which they are composed. Figure 6.4 shows a fern. It has an overall shape that is echoed by each of its fronds; these are in turn composed of subfronds that again have the same shape.

Figure 6.5 shows a computer simulation of a fern pattern based on the iteration of the basic frond shape: there is no doubt that fractals can make good models of plants of this type, which are themselves highly self-similar. The particular method used in making the fern model is called an iterated function system, which works by prescribing a set of scaling and replacement rules. Fractal patterns made in this way have several properties that are of interest for biology. One is self-similarity: if you enlarge the pattern, it will have the same overall appearance at the enlarged scale as it did to begin with. An example of this, which we saw in chapter 1, is a coastline like that of Britain, and this was the example Mandelbrot considered in formulating the idea of fractals. From an airplane you see much the same jagged outline as you would see looking down on rocks from the top of a cliff or standing looking into a rock pool. Figure 6.6 shows a computer-generated model of a fractal that resembles a coastline at different levels of magnification. Although the detail is different in each frame, the overall properties of the frames are the same—they have the same degree of roughness, or fractal dimension.

Another striking property of fractals is their complexity, which is so great that it is capable of rivaling that of life itself. We have already seen how new information is created by the process of iteration, and life is very rich in information content. Fractals are the most plausible way of representing this informational richness and complexity.

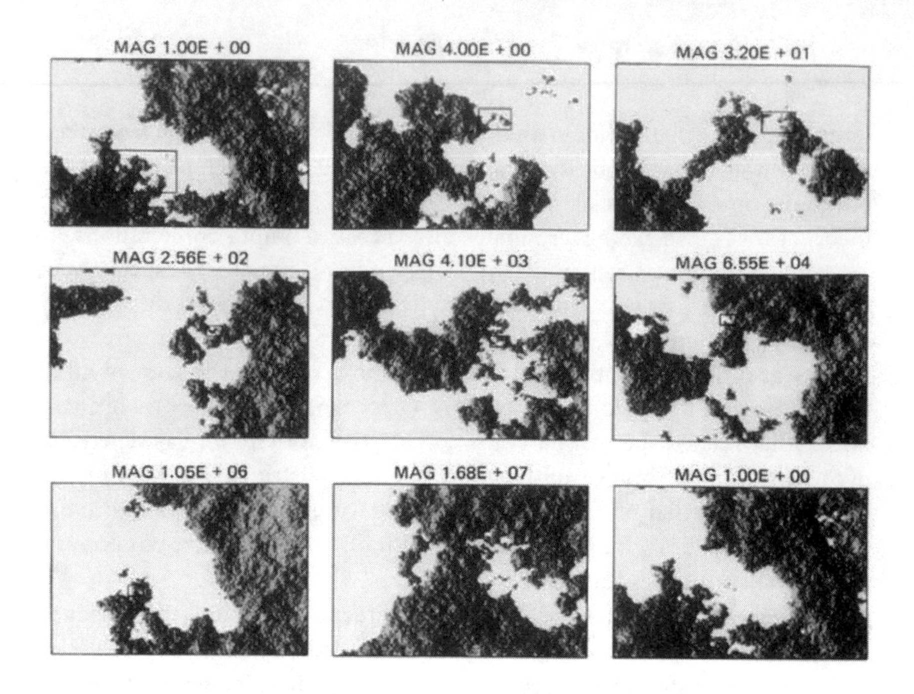

FIGURE 6.6 Computer-generated model of a fractal that resembles a coastline. [From H-O Peitge and D. Saupe, eds., *The Science of Fractal Images* (New York: Springer-Verlag, 1988).]

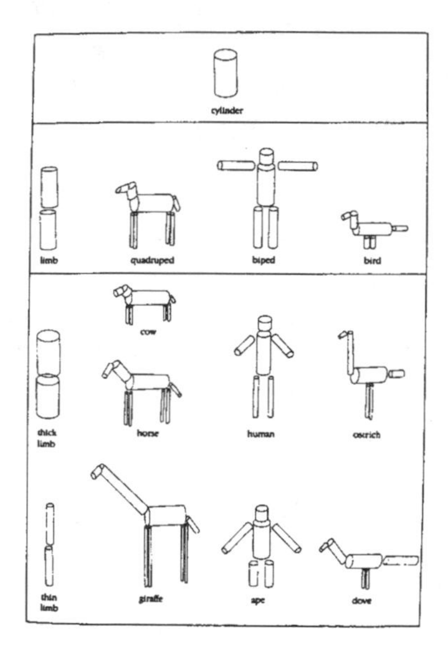

FIGURE 6.7 Model description of quadrupeds, bipeds, and birds. [Reprinted from D. Marr and H. K. Nishihara, "Representation and Recognition of the Spatial Organization of Three-Dimensional Shapes," *Proceedings of the Royal Society of London* 200 (1978): 269–294, with permission.]

It is hard to specify how much information an organic form such as the human body contains, partly because there is no adequate way yet of measuring morphological or structural information in biology, but it is clearly very large. Simulations of organic shapes are sometimes produced using a few simple Euclidean forms such as cones and cylinders, but these usually produce something that is not adequate as a realistic portrayal of the human form. Figure 6.7 is an example taken from the work of two perceptual psychologists,[4] showing models of birds, quadrupeds, and bipeds. In their method a body is decomposed into a hierarchy of cylinders, one for the trunk, others for legs or arms, and so on. As can be seen, the effect lacks realism. Most of these modeling approaches severely distort reality—the bodies are nothing like this—and an acid test is the effect they produce on the viewer when they are used for making cartoons.

The inadequacy of such models might seem discouraging to analytical attempts to simulate the forms of organisms. But these limitations can be transcended by the use of fractals, by means of which it is possible to produce literally endless detail from the iteration of quite a simple formula. There is in fractals the potential for producing the amount of information required from a very economical set of assumptions. This complexity that iteration is capable of producing is what makes chaotic processes and fractal forms the most likely candidates for modeling living shapes.

THE MANDELBROT SET—A MODERN HORN OF PLENTY

The fractal patterns we have examined so far, such as the snowflake curve, are produced by iterating simple linear functions. However, even more complex fractal patterns may be made if the function iterated is a nonlinear one. The well-known Mandelbrot set, and its related Julia sets, are fractals of this type and it is to them that I wish to turn as the theoretical foundation of the geometry of life. I shall therefore spend a little time on the Mandelbrot set (abbreviated to M) and its variations.

The set M is produced by iterating a very simple formula involving real and imaginary parts that correspond to the x and y axes.[5] A picture of M for 150 iterations is shown in figure 6.8. Some features of its shape are immediately striking. The black inner area has a very complicated border, with many elaborate curves and cusps, with processes sticking out of it and fissures cutting into it, but it also has a very distinct boundary within which these features are contained. Thus, while the big infold on the right-hand side is studded and pitted with irregularities, it still has a clear outline, known as a cardioid or heart-

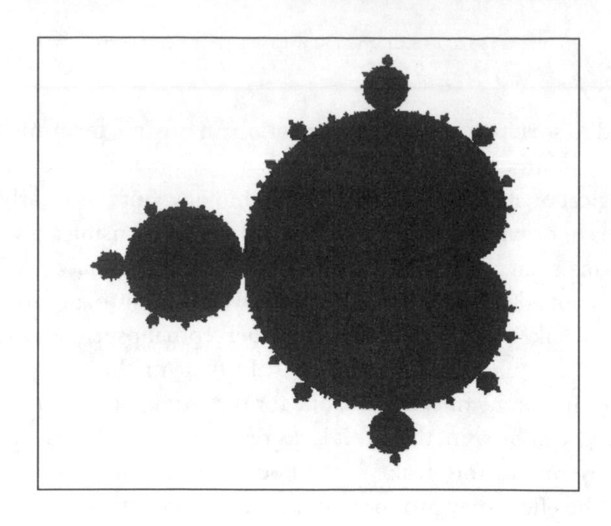

FIGURE 6.8 A picture of M for 150 iterations.

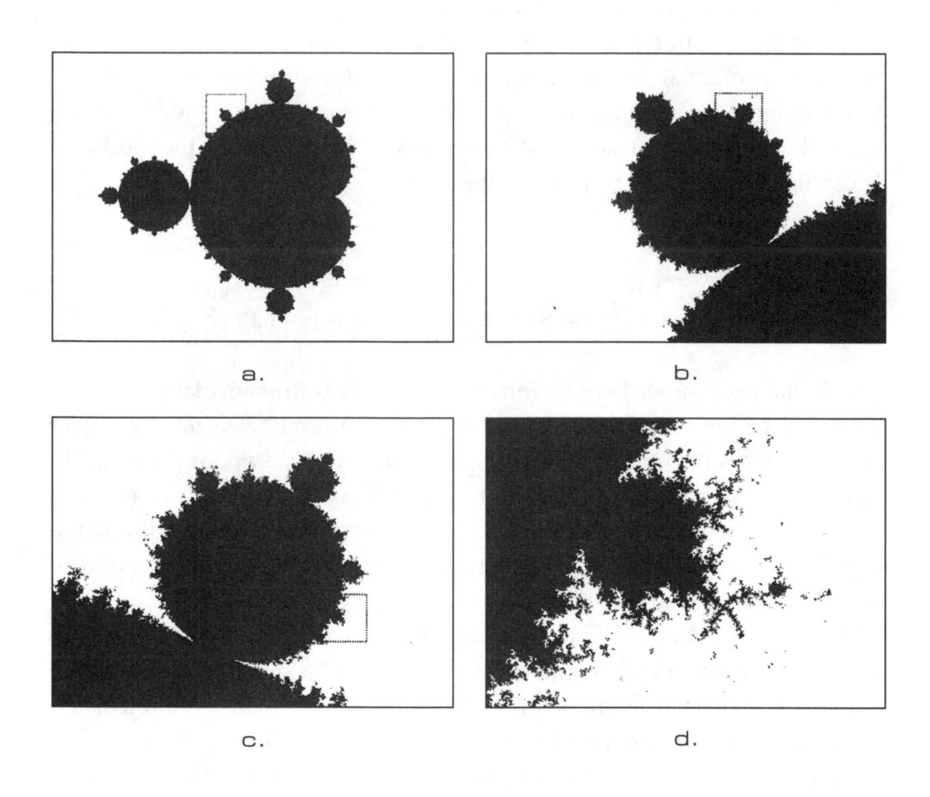

FIGURE 6.9 Successive enlargements of M.

shaped curve, to which they adhere and which they ornament. This is the largest area of M. The next largest part of M is circular and lies to the left of the cardioid, to which it is joined by a narrow isthmus left by the infolds from above and below.

An important feature of M is that, because it is self-similar, its boundary reveals more detail as the number of iterations increases, and this detail can be seen on computer enlargements of the set. To enlarge M we do not blow it up in the photographic sense; if we did this, we would lose picture definition quite quickly. What is done is that a small part of the set M is computed for a larger number of pixels on the computer screen, and/or for a greater number of iterations. Figure 6.9 shows what happens when you enlarge M in this way. The frames show M and three successive enlargements [(b)–(d)] of the part of M enclosed in a box in the previous frame. This enlargement can be continued indefinitely, but as you go on with the process, the number of iterations has to be increased in order to produce more detail, both because each computed fractal is finite and because of the limitations of the precision of the number representation used in the computation. As the enlargement process continues, fantastic and highly complex shapes emerge. Overall, the impression is one of amazing complexity—so much so that the scientists who first produced these shapes on their computers thought they had made some kind of error. A purely mathematical object of such complexity was quite unknown.

COMPLEXITY FROM SIMPLICITY

What was particularly hard to believe at first was that such a complicated object could be produced as the result of such a simple formula. Indeed, the formula is about as simple as it could be and the mystery is why its associated pattern (called an escape diagram) had not been discovered long ago. The answer, as with so many other things, lies in iteration. It is iteration that produces complexity out of a simple process—in this case a formula. After only four iterations of the formula we have produced a curve whose equation involves the sixteenth power of the variables. This is well beyond the ability of anyone to analyze by mathematical methods. Such complexity also seems well beyond the powers of the unaided individual to even calculate and thus had to await the invention of the computer. Without computers, it would have been possible to produce shapes like these only by means of tremendously long, drawn-out calculations, so lengthy as to effectively deter anyone from pursuing such an arduous and unpromising path.[6]

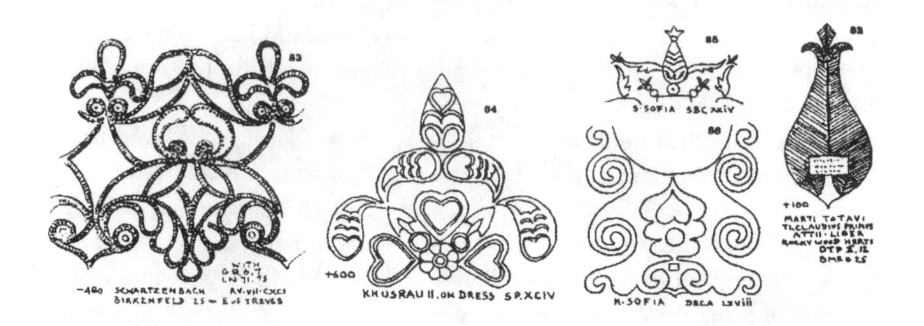

FIGURE 6.10 Flinders Petrie's pictures' affinities to variants of M. [Reprinted from F. Petrie, *3000 Decorative Patterns of the Ancient World* (New York: Dover, 1986), with permission.]

FIGURE 6.11 "BuddhaBrot" figure, from original by Melinda Green.

Nevertheless, it is tempting to wonder whether in some culture of antiquity the existence of M was suspected, even if it could not be computed in detail. It is known as a result of the work of Derek J. de Solla Price[7] that the Greeks had, as long ago as 80 B.C., a digital computer (the Antikythera Mechanism) that was used for calendrical calculations. The Antikythera Mechanism used gears, just as Charles Babbage's nineteenth-century Difference Engine did. Could such a device ever have computed the value of squared complex formulas sufficiently well to visualize the main features of M? This seems unlikely, if only because the idea of complex numbers did not enter mathematics until their discovery by Cardin in 1545.

Is it possible that the existence of M was known in former times in some other way? Although early arithmetic systems did not incorporate the idea of complex numbers, work on sequences, especially the Fibonacci sequence, which corresponds very well to the morphology of plants, had already foreshadowed modern use of mathematics in biology. The golden section was also know from classical times (although apparently its relation to the Fibonacci series was not suspected). Certainly the shape of the Mandelbrot set has acquired the status of an icon in today's world. Perhaps by virtue of its universality it may have obtruded into the psyche in former times also as a sort of Jungian archetype. There are many shapes in iconography that resemble parts of M. Flinders Petrie's *Decorative Patterns of the Ancient World*[8] contains a wealth of designs that show marked affinities to variants of M, and some of these are shown in figure 6.10.

Sometimes quite complex designs show resemblances to the set M. In religious iconography, for example, the form of the seated Buddha figure is strikingly echoed in a modified form of M created by Melinda Green[9] (see figure 6.11). Buddhist and Hindu meditative experience may well have led to fractal visions and dreams, as may visions from other cultures, especially those exploring drug states, including the synthetically induced "psychedelic trips" of today. Could this universal use of pattern and design, sometimes in quite widely separated cultures and times, have arisen unless there was an abiding basis for it in the human psyche?[10] Certainly the form of M confirms the ancient doctrine that tells us that the "Many flow out of, as well as into, the One."[11]

SYMMETRY AND BRANCHING IN NATURE

The first and perhaps most surprising feature of M that makes it a candidate model for life is that it is continuous. Although there appear to be places where the central area is cut off from its tendrils, it can be shown mathematically that,

FIGURE 6.12 Birch tree.

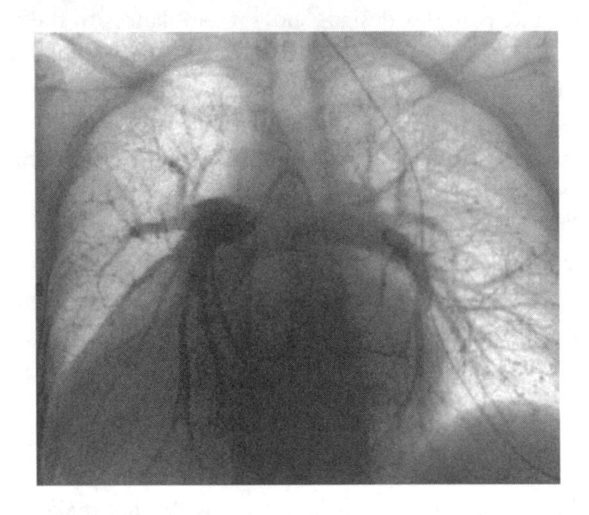

FIGURE 6.13 Pulmonary angiogram.
By kind permission of Professor Jeremy Ward.

no matter how many times we iterate, the areas are all joined together. This is a primary requirement for an organism; each part must be connected to the whole, for nutrition, information flow, and control.

Another important feature of M is its symmetry, in this case about one axis—the horizontal axis. This characteristic is shared by living forms: almost all advanced life has symmetry of some kind. The human body displays a high (but not quite perfect) degree of left/right symmetry, and in the shapes of plants symmetry is still more evident. Complicated symmetrical patterns appear in flowers and their component parts, often involving several axes. In advanced animals such as vertebrates the symmetry is usually about a plane running the length of the body from head to tail (the anterior–posterior axis.) It is true that this symmetry is not exact and there is a degree of asymmetry in most species. In humans we notice this in the form of dominance for a particular faculty such as handedness, or in some cases the advantage that one eye or one ear has over the other. However, basic symmetry is a universal feature of advanced living things and is also one of the features offered by fractal models.

Higher organisms share another feature that is of great importance—they all contain branching structures. These appear in different ways; sometimes the body plan is dendritic. A tree, for example, is a classic dendrite (dendritic means treelike), composed of a central trunk with branches, these branches subdividing repeatedly, as in the birch tree in figure 6.12

Sometimes it is the internal structure that is dendritic, such as the respiratory system. Figure 6.13 is an angiogram showing the branching of the human lung. It is very similar in shape to the overall envelope of a tree such as a birch or an oak. Lungs are paired, which emphasizes their symmetry in humans. The pulmonary system has been found to have fractal characteristics.[12] There are other, more systematic resemblances between tree and lung: the total cross-section of the alveoli remains approximately constant at each level and the total cross-section of the branches plus the trunk in trees is also constant at any particular height.

It is an interesting sidelight to note that much of mythology is based on the tree, its form and its magical function, suggesting that there is a fundamental and deep realization in the human psyche of the significance of this form for human—indeed, for all—life. The tree enters into the mystical tradition of many religions—notably Christian and Buddhist. In his study of mythology, *The Golden Bough,*[13] James Fraser stresses the significance of the tree in myths from widely separated cultures. It is clear that the dendritic structure of living forms has a significance for most if not all civilizations.

The nervous and circulatory systems of vertebrates also have this dendritic structure. Figure 6.14 is the outline of the human nervous system as seen from

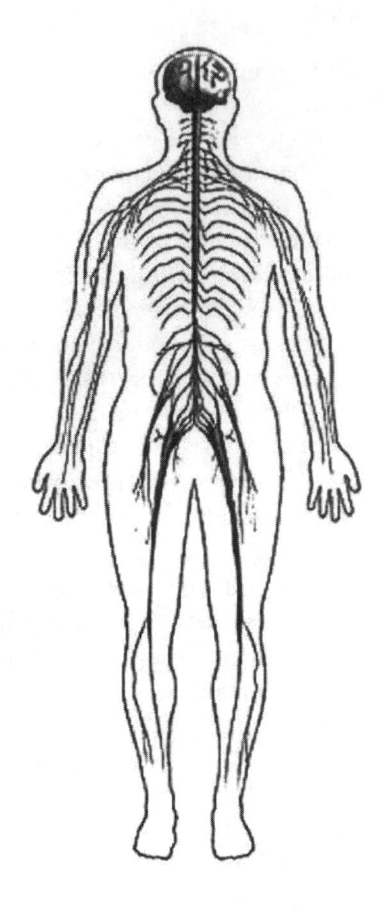

FIGURE 6.14
The human nervous system, as seen from the front.

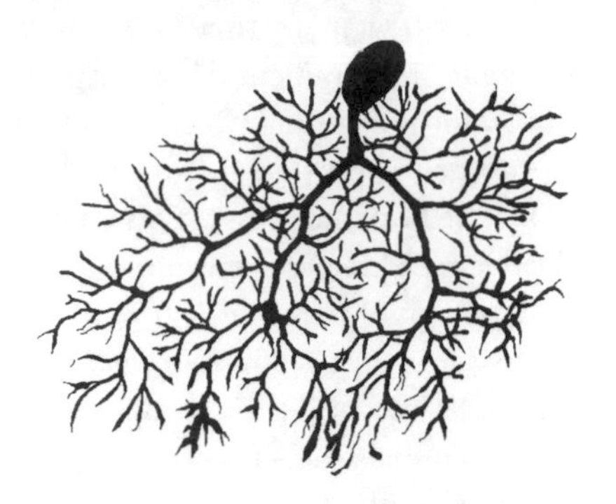

FIGURE 6.15
Purkinje cell. By kind permission of Michael Häusser and Arnd Roth, University College, London.

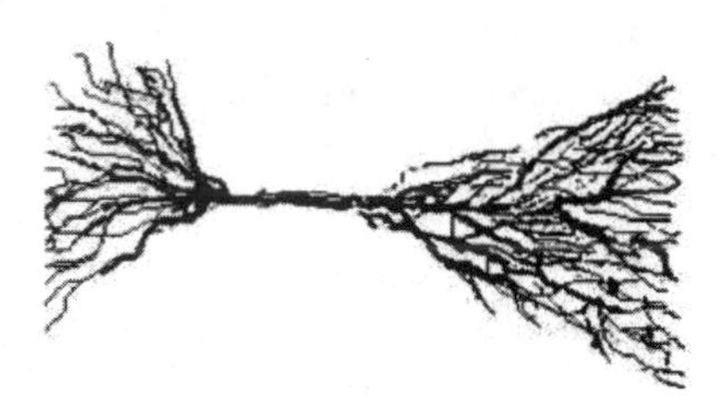

FIGURE 6.16 Neuron.

the rear. It shows quite clearly that the nervous system has a dendritic form and it is repeated on a smaller scale: figure 6.15 is a drawing taken from a gold stain of a brain cell called a Purkinje cell. Each neuron is also a combination of two trees, the dendritic and the synaptic, with thousands of branches. The basic structure of an interneuron is shown in figure 6.16. Figure 6.17 shows the human circulatory system, which again shows a dendritic structure.

The body plan of most eukaryotes (multicellular organisms) is both symmetrical and dendritic; that is, it branches repeatedly in a symmetrical fashion.

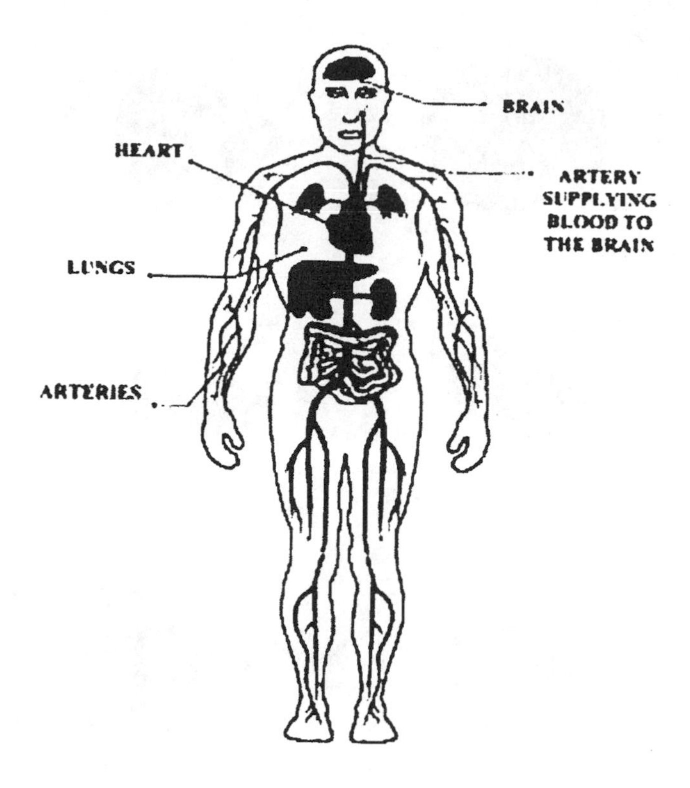

FIGURE 6.17 The human circulatory system.

In animals the subdivision of the body ends after three or four repetitions; in plants it may be repeated more than half a dozen times, typically seven in many species of trees.

THE PICTURES IN THE FIRE

The resemblances between one organism and another amount to an apparently common body plan over a very wide range; this is sometimes casually dismissed, as if what is obvious is of no importance. Or it is said that shapes may look similar for no better reason than that we read similarity into them; that they are "pictures in the fire"—the chimerical resemblance of shape that arises, not from a real structural similarity, but from a kind of perceptual illusion.

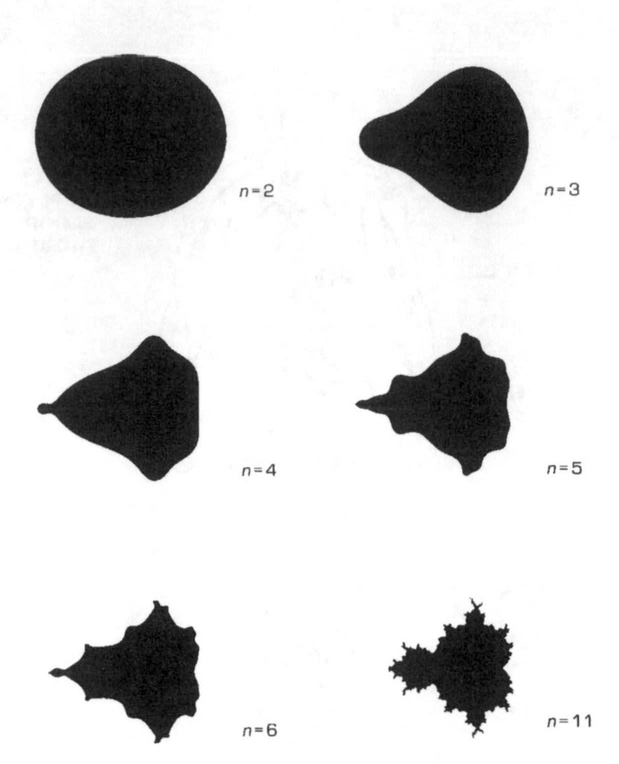

FIGURE 6.18 Shapes produced in M when lower iteration numbers are used.

It is an interesting point in itself why the human perceptual system sometimes sees ambiguous patterns as having a distinct and familiar shape, but there comes a point at which resemblances cannot be so lightly dismissed. The similarities of all these instances of symmetry and dendritic structure in biology make it inevitable that the question be asked: is there not some principle of organization that applies at many different scales and in many different places that has given rise to such similar shapes? What have the tree and the neuron in common that they should have such a remarkable similarity of form? It is only by studying the constant factors between situations that we learn about the forces outside those situations that shape them all: it is not the falling of one object that is suggestive of gravitation, it is the falling of all objects. If things have the same form, then must they not have been shaped by the same processes? And surely this law applies to living things just as much as to inanimate objects. If there is such a remarkable universality (almost identity) of

FIGURE 6.19 Enlargement of M for $n = 50$, colored alternately black and white.

FIGURE 6.20 A photograph taken looking up through an oak tree.

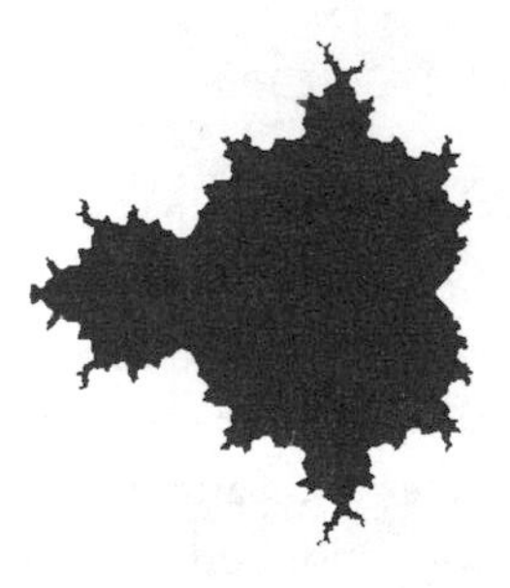

FIGURE 6.21 M for $n = 20$, showing the root system.

form among many types of biological structures in nature, it can only be as a result of the same processes operating in each and every case.

When the processes of biology are seen as iterative, then much of what was previously obscure suddenly becomes clear. Parts of fractals produced by nonlinear iteration resemble organic forms. To see this, it is only necessary to draw the set M in a slightly different way. Part of the definition of M is that it is iterated a particular number of times; we then color everything inside the set[14] differently (in this case black) to distinguish it from what lies outside the set. Figure 6.8 was drawn in this way using 150 iterations ($n = 150$). If we draw the boundaries of the set for lower values, say $n = 1$, $n = 2$, $n = 3$, _, we get some interesting shapes. Figure 6.18 shows the shapes produced when a series of lower iteration numbers are used. For $n = 1$ (not shown) we get a circle, one of the most basic shapes in biology. For $n = 2$ we get an ellipse, also a common form. For $n = 3$ we get a pear shape. At $n = 4$ the shape produced is recognizably mammalian. For $n = 5$ it resembles a ray or skate, and $n = 6$ is like the outline of a bat. At around $n = 12$ the shape is rather like a tree trunk cut through just above the roots.

In figure 6.19 we see an enlargement of a different part of the set. The regions corresponding to successive iteration numbers have been colored alternately black and white. It is rather like looking up into a tree, seeing the branches, twigs, and leaves. Figure 6.20 is a photograph looking up into an oak tree; the resemblance in form between this and figure 6.19 is striking. Figure 6.21 shows M for iteration number $n = 20$, which appears rather like a root system.

Are these just pictures in the fire, or are they not rather pictures that are formed by the fire of life? The shapes produced as parts of the set M and its derivatives are certainly highly suggestive of organic shapes. If it were possible to show how organisms could develop in such a way as to take such forms, then we might be well on the way to understanding morphology.

PATTERN FORMATION BY BIFURCATION

First, let us state a very general point about iterative processes in development and development in general: the shape of an organism is largely determined by the pattern of division of its cells (although changes in cell size are also important). As the organism grows, it also develops, starting out as a simple shape and becoming more and more complex and patterned as it grows. There are at least three reasons for this: cells divide more often in some parts of the organism than others; cells migrate from one part of the body to another; and (importantly) some cells die off during development in a process known as *apoptosis* or programmed cell death. An example of the last process is the webbed skin between the fingers of an unborn human child, which disappears while it is still an embryo. Between them, these processes give an organism the shape it has when fully grown.

In some simple organisms the cells divide according to their lineage, that is, which cell they are descended from, so that development is completely programmed or autonomous. In the fruit fly *Drosophila* on the other hand, development is specified by the position of the cell within the body, which is a very different kind of program. In either case the form of the full-grown adult is specified by the same processes of cell division, migration, and cell death. The tiny nematode worm *Caenorhabditis elegans* (usually abbreviated to *C. elegans*) has 959 cells, of which 302 are neurons. Everything is known about the way development of *C. elegans* occurs because it is transparent. Here the development is regulated by a mixed system—partly by the lineage of the cell and partly by the position of the cell within the organism—and this is more characteristic of the developmental program of higher animals. In the embryo stage of development the worm is compartmentalized, with patterns emerging from each separate section.

In the language of dynamic systems a division or splitting is known as a bifurcation. The decision of a cell to divide or not is an example of such a bifurcation, and is dependent on a state of chemical nonequilibrium, which in turn is probably a function of the length of particular complex molecules such as the nucleic acids—DNA and RNA. The length of the relevant molecules is further likely to be determined by nonlinear processes taking place during the manipulation of nucleic acids. So the pattern of growth is likely to be related to the outcome of bifurcation processes. What remains to be worked out (although it is a very great deal) is how these processes determine the shape of living things and by what particular formulas.

There is more specific evidence, however, that supports the idea that the shapes of living things result from such dynamic bifurcating processes. As already pointed out, the body plan of higher-order organisms is dendritic, a feature characteristic of certain fractals. Furthermore, many organic forms have a measurable fractal dimension. West and his colleagues[15] have found a fractal scaling law for a wide range of biological organisms, covering a span of about ten orders of magnitude, which follows a 3/4 scaling exponent.

Sections of the Mandelbrot set and the Julia sets also resemble organic forms. The largest area of M is a cardioid, a shape common to the violet leaf, the kidney, and the heart itself. Spirals can be found all over M,[16] especially in the infold that separates the two largest areas. All sorts of flowers, from sunflowers to cauliflowers, show spirals in their florets, and parts of many plants,

FIGURE 6.22 The spiral structure of *Nautilus*.
By kind permission of Mr. William Fecych and the MIT Nuclear Reactor Laboratory.

FIGURE 6.23 Shape resembling a human leg.

such as spruce and pine cones and the pineapple, grow from the stem in a spiral when viewed from above. In the animal kingdom, mollusk shells have a spiral structure, notably the *Nautilus* (figure 6.22), as have the horns of the sheep and goats and even human finger nails (when grown long enough). The human cochlea is spiral in structure (cochlea is Latin for cockle). The outline of the buttock pursues a spiral path; the Finnish word for buttock is *pylly*,[17] derived from the Greek *phyllum*, meaning "leaf."

Sometimes the resemblances between M and organic forms can be even more striking. Shapes like sea-horses are found on either side of the large infold. Figure 6.23 shows a shape something like a human leg; this is found in the region of M that is at the bottom of the main cardioid and it appears at around the twelfth iterate.

FRACTALS—FINITE AND INFINITE

Mathematically, fractals are often thought of as being iterated an infinite number of times, but in the natural world this is unrealistic. No real-world process can continue for ever and a computer process cannot be executed an infinite number of times. When we look at pictures of M, what we usually see is something that has been iterated perhaps 500 or 1000 times; it is not practicable to do more on most desk-top computers. A time limit (or iteration limit) has to be set on the computation. So all *real* fractals, as opposed to mathematically idealized fractals, are finite.

In one characteristic, finite fractals have the same property as infinite ones: they have a fractal dimension that is greater than their Euclidean dimension.[18] The only figures for which this is not true are those composed entirely of straight lines. A circle is the simplest nonlinear shape, but even a circle is fractal: if you measure a circle with a ruler, its diameter (Euclidean dimension) is constant but the measured length of its circumference (fractal dimension) will increase as the length of the ruler decreases, over a certain scale. *Any* nonlinear figure is a fractal of some sort. We can thus begin to appreciate how ubiquitous fractals are in natural shapes: all are finite fractals; the simpler the curve, the fewer the number of iterations required to produce it.

POSITIONAL INFORMATION AND ITERATION

Regular fractals of the kind we are talking about are the outcome of iteration. Each step of the cell cycle is iterative and when the cell divides, new informa-

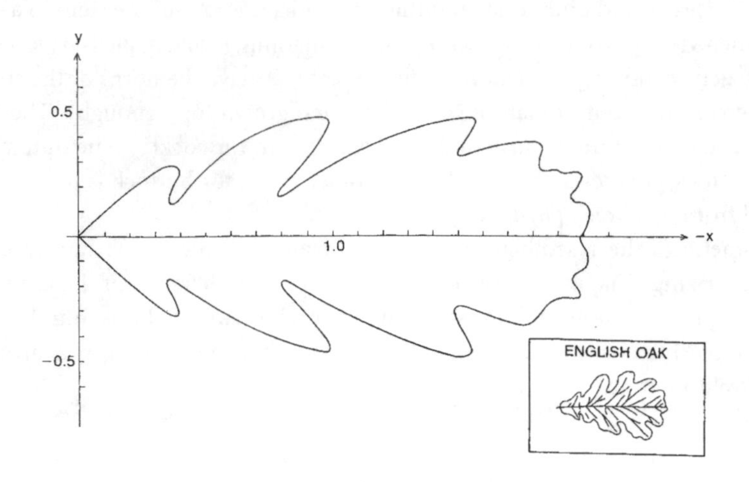

FIGURE 6.24 Oak leaf outline, produced by iterating a formula of less than 50 symbols. [From R. J. Bird and F. Hoyle, "On the Shapes of Leaves," *Journal of Morphology* 219 (1994): 225–241.]

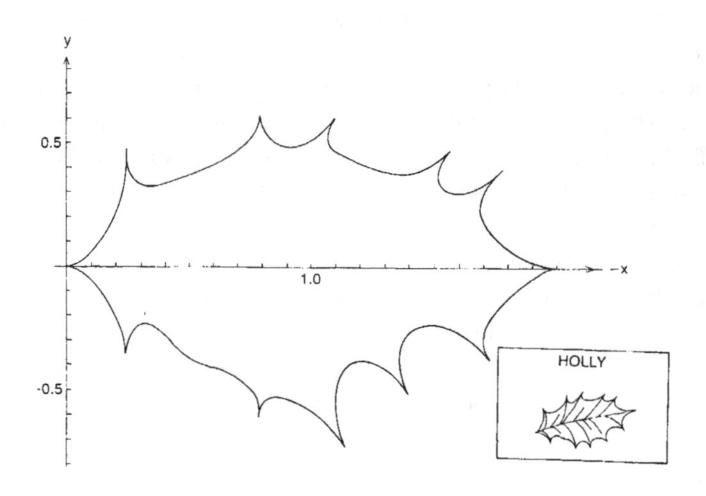

FIGURE 6.25 Holly leaf outline, produced by iterating a formula of less than 150 symbols. [From R. J. Bird and F. Hoyle, "On the Shapes of Leaves," *Journal of Morphology* 219 (1994): 225–241.]

tion is carried forward to the next cycle. There is a tremendous amount of information packed into a cell, but let us focus on one thing—information about its position. It is a cardinal principle of modern theories of development that a cell knows where it is within an organism, probably relative to the "mother cell" from which it originally sprang.[19] The cell keeps a record of this position, and also other information such as its orientation, and it is updated each time the cell divides.[20] This positional information is probably held in protein representation.[21]

So the cell knows one thing that is essential for the production of fractal shape by iteration—it knows its position.

The other component of the production of fractals by iteration is the process that is to be iterated. In the case of the Mandelbrot set, for example, this is the formula $Z' = Z^2 + C$, where Z' is the new value if the complex variable Z.

If Z stays less than a certain size under iteration, it remains in the Mandelbrot set. If Z gets large enough, the point escapes from the Mandelbrot set. Developmental biologists draw diagrams of the cells in a developing organism showing their final destination after they have undergone division and specialization and reached full function within the organism. This is called a *fate map* of the cells. In the same way, the Mandelbrot set is the fate map of the points, showing what will happen to them when they have been iterated enough times. The difference between M and a fate map is that the cells move but the points remain static.

The suggestion Fred Hoyle and I made is that the shapes of many leaves can be explained by the iteration of a simple nonlinear equation based on the set M. To make this suggestion plausible, we considered the shapes of leaves and showed how they might be constructed. Figure 6.24 is drawn following a rule from our paper[22] and shows an oak leaf, together with a drawing of a natural leaf (inset). Figure 6.25 shows a holly leaf outline and drawing. It is a fairly convincing representation of the real thing, even capturing the slight bulge between two lower spikes where a new spike is going to appear at a later stage of development. Both figures were produced by iterating formulas[23] of less than 150 symbols.

WHY IS NATURE FRACTAL?

Why should living things be fractal? Given that it is a frequent aspect of natural structures, why is it also a useful one? One answer is that in certain systems fractality is an adaptive form. West and his colleagues[24] have demonstrated

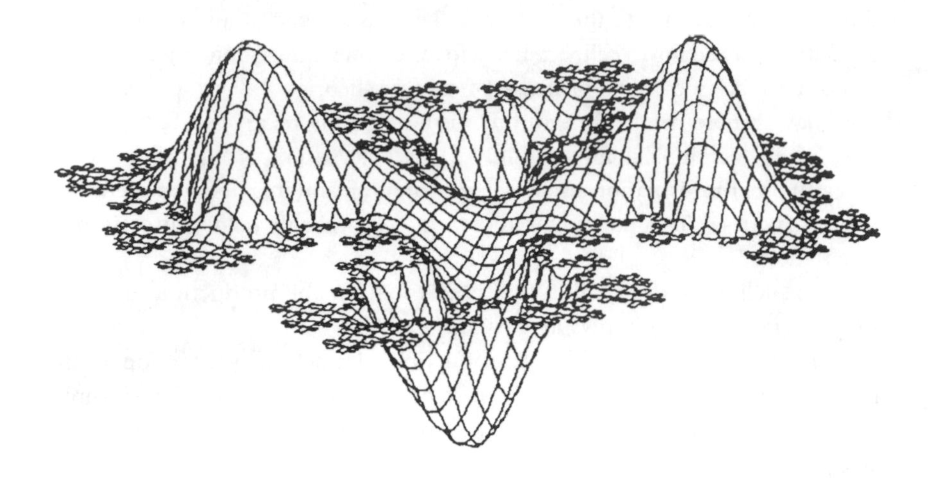

FIGURE 6.26 The dissipation of shock by a fractal structure. [Reprinted from B. Sapoval, T. Gobron, and A. Margolina, "Vibrations of Fractal Drums," *Physical Review Letters* 67 (1991): 2974–2977.]

that a universal power law governs the changing size of organisms during development. The underlying reason for this is the balanced requirements of consumption and distribution of resources for the energy requirements of growing cells. A fractal capillary system for blood circulation is optimal for distribution, which grows according to the index 4/3, while consumption grows linearly: when demand outstrips supply, the organism stops growing. Another reason is the sheer power output of the heart in large organisms. Sapoval[25] showed how a fractal shape for the circulatory system lessens the shock of the pumping heart. Figure 6.26 illustrates the principle of dissipation of shock by a fractal structure.

The nervous system may similarly be fractal for the optimal dissemination of information carried by nerve impulses. We have also seen how living things are maintained in a state of moderate chaos. It is one of the messages of this book that chaos and fractals are examples of iterative processes and that life, too, is characterized by these processes.

Another advantage of fractal form is its symmetry; as we have seen, symmetry is a feature of both advanced organic form and many fractals. Fractals also have the most detail where it is needed—at the periphery: a structure based on fractal form ensures an even distribution of air, blood, and other nutrients.

A specific advantage of M is its connectedness; the parts of the set are all joined, and even when it looks as though they might be separate, a thin filament runs through the isthmus that threatens to divide them. The biological need for connectedness between parts might seem like a truism, but organisms must remain connected in order to survive and maintain both their individuality and their organic integrity, and this need is met by the connectedness of M. This connectedness applies to internally communicating systems such as blood vessels and nerves, and also to the adaptive connectedness of the envelope of the organism: examples are the skin and fur of humans and animals.

Is this, then, the basic pattern of life—for that would not be too grandiose a description of it—a Form that can be transformed into a library of forms by simple changes in the formula to be iterated during cell growth, so that the organism develops to take up its fully grown shape? The resemblances of pattern between living things would then be the inevitable consequence of the mathematics of growth: organisms that grow using cell division and in doing so execute some variation on this basic arithmetic theme of life. The mechanism they may use to do so will be the subject of the next chapter.

7

THE LIVING COMPUTER

An unapparent connexion is stronger than an apparent.
(FRAGMENT 54)

THE BIOLOGICAL COMPUTER

The idea I am suggesting is that the developing organism is capable of working out mathematically what shape it "ought" to be and using the results of those calculations to control cell growth; in other words, that there is a biological computer in each cell in your body and in that of every other living thing. Because this computer is constantly at work in every cell, the organism is a massive parallel processor with a power as yet undreamed of in the laboratories of IBM or Intel. If this idea is correct, then even a simple organism is a computer many millions of times more powerful than any device, either current or capable of being built in the foreseeable future.

To explain why I think this is so, I want first to set out a possible mechanism for the storage of information in this biological computer. The idea is that repeat sequences in the DNA, which are abundant throughout the genome, are the basis of computations leading to morphology and other aspects of gene organization. I will try to show that it is highly likely that these repeat sequences represent numbers and that the computing or number-crunching is done by means of DNA and RNA processing. The first question must be: what is the evidence for the representation of numbers in such DNA strings?

It is a cardinal point that any representational system is more flexible if several different alternative ways can be used to represent the same message. In nature as in human inventions this principle is universal. For instance, video signals can be stored on film, disk, or tape, or transmitted as waves. In genetic coding there are several different triplets that are translated into the same amino acid.[1] For example, CTA, CTG, CTC, and CTT all code for the amino acid leucine, and so do the sequences TTG and TTA. (The four bases used in DNA—adenine, thymine, cytosine, and guanine—are abbreviated A, T, C, and G.) Three different triplets indicate a stop signal for terminating translation. It is worth bearing in mind, as we search for possible DNA and RNA codes, that nature has often found alternative ways of transmitting the same vital message.

REPEAT SEQUENCES IN DNA

It is not difficult to locate segments of nucleotide chains that might fulfill the function of encoded digits. Repeat nucleotide sequences are a ubiquitous feature of the eukaryotic genome.[2] Repetitive DNA is found in organisms ranging in their degree of complexity from simple yeasts to humans.

DNA has an already known coding function by which it is translated into protein; this is known as the universal genetic code. In the genetic code a group of three bases, known as a *codon*, specifies the production of a particular amino acid. People sometimes suppose that this accounts for the whole of the structure and function of DNA, but this is not the case. There are long sequences of DNA (sometimes called *silent* or *junk* DNA) for which no function is known. In humans, for example, it accounts for over 95 percent of what is known as the genome (perhaps a misnomer in view of the sparseness of the genes themselves). Many of these repeated sets of nucleotides, repeat sequences or repetitive DNA, are part of this silent or noncoding region of the genome. This repeat DNA is found at various sites on the genome, particularly at the centromeres and telomeres of the plant or animal chromosomes, although repeat sequences appear at other sites also. (The centromere is the junction of the two main parts of the H-shaped chromosome; the telomere is the part most distant from the junction.)

Various attempts have been made to identify the function of these DNA repeats, but so far with limited success (except in one area, which we will concentrate on shortly). It may be that the repeats have a number of functions in the organism: they may be involved in gene silencing or they may be a mechanism for producing genotype variation.[3] Some are the remains of sections of the genome transposed from one place to another during evolution (transposons). It has even been suggested that repeats serve no useful function, being merely passengers carried by the rest of the genome that have not experienced enough selective pressure to lead to their extinction.

Some molecular biologists suppose that the structure of DNA is dependent on natural selection, as are the characteristics of the individual animal or plant;

this theory is known as molecular evolution. Indeed, a whole theory of evolution is based on the idea that it is components of the genome rather than the individual organisms of a species that are the units of natural selection, a view put forward in Richard Dawkins' book *The Selfish Gene.*[4] The ubiquitous presence of repeating molecular chains in the DNA in such huge numbers should therefore imply a function; but no general function has yet been found. The hypothesis I am now suggesting is that the repeat DNA is the basis of computations leading to morphology and other aspects of gene organization.

THE VARIETIES OF REPEAT SEQUENCES

There are two main kinds of such sequences, highly repetitive and moderately repetitive. The highly repetitive sequences are not transcribed into RNA and cannot convey information between cells during cell division. The number of repeated units (the *repeat number*) varies from 2 to millions and the size of the units, or *repeat length*, also varies widely. At the upper end of the repeat length scale, whole genes are repeated, and are functional. However, very long repeats known as *pseudogenes* (repeated but nonfunctional gene-like sequences) are also found. At the lowest end of the repeat length scale, a single base may occur with unexpected frequency in certain parts of the genome. Pairs, triplets, and other multiples are also found.

Repeats may be classified into a number of types according to their structure. Repeats are called *direct* if they are separated by nonrepeating stretches; when they are adjacent to one another they are called *tandem direct. Inverted* repeats are usually associated with structural stems of DNA leading to loops, where the joined arms of the hairpin mean that a complementary *palindromic* sequence of bases appears on the opposite arm, the limiting case being a hairpin that has only one other, unlinked, base at the end of the stem. In these hairpins, continuous palindromes occur in which the series of bases is accompanied by the reverse sequence on the other strand. *True palindromes* are said to occur when the sequence is followed by its reverse on the same, unlinked strand. Within the longer repeat units, which may be of the order of several kilobases, shorter repeat sequences may appear, giving a nested structure of repeats. The depth of this nesting may be as many as 4 levels with repeats of 2 to 5 base pairs as the smallest repeat unit and repeated genes the largest.

There are two types of eukaryotic DNA, called *Xenopus* and *Drosophila* after the representative species that carry them. It is known that Xenopus-type DNA contains large sections of repetitive material, with reiterated sequences of thousands of codons separated by intermediate spacing material. Perhaps 10 percent of this type of DNA consists of highly repetitive sequences, with a total of around 25 to 30 percent of DNA being either repetitive or highly repetitive, depending upon species. The average of the most frequently occurring lengths of these repetitive sequences is around 300 nucleotides. Most of these repeats occur in the regions known as *introns*, which are only exceptionally transcribed into RNA (as opposed to the *exons*, which are transcribed). Because these intron sequences are not transcribed, it was thought that they are of no use as genetic information. However, they could be used for decisions such as whether the cell should divide or not, or whether or not to express a given gene product. Their nontranscription into genes then becomes more explicable: they have another function to perform, one that might be thought of as executive rather than administrative.

In *Drosophila*-type DNA the shorter repetitive sequences of *Xenopus* DNA are lacking, and instead there are longer repetitive sequences of about 6,000 nucleotides. Since morphologically similar organisms can have different DNA types, it has been argued from this finding that repetitive sequences in DNA can have no developmental significance. However, if the thesis advanced here is correct, and if repetitive sequences of nucleotides encode digits that can be interpreted by a computing mechanism in the cell, then it must be apparent that the particular representation used for a digit (say a 1 or a 0) may differ in different organisms, provided the computing mechanism can interpret the code correctly. Just as a sound signal may be equivalently represented in many media, so two different genetic coding mechanisms may represent the same information. The dual nature of DNA found in closely related species in no way weakens the case for a genetic computation mechanism; indeed it strengthens it by emphasizing the syntactic nature of such a code, if it exists. This is an instance of the principle mentioned at the start: that nature may have adopted different representations of the same information.

Since the sequencing of the mouse and human genomes, interesting comparisons have emerged, and some of these are in the area of repeats. In both mice and humans, repeat sequences form a substantial part of the genome. A large number of these are transposons. These are a kind of repeat element resulting from sequences that have been copied from their original site to another part of the genome. It is thought that they are there because of evolutionary changes that can lead to copying out a gene many times. Four kinds of transposons are of current interest: LINEs (long interspersed nucleotide elements, of up to 6,000 base pairs); SINEs (short interspersed nucleotide elements, of around 300 base pairs); LTRs (retrovirus-like elements with long terminal repeats, called retrotransposons, which reproduce via RNA); and finally, DNA transposons, which reproduce within DNA itself.

REPEAT SEQUENCES AND INFORMATION CODING

The repeat sequences of DNA would be ideal as a means of representing information. The big question is, what sort of code might they use? The most likely code is unary, or counting in ones. This is the simplest possible arithmetic system and therefore the one that is most likely to have arisen during the course of evolution. In unary arithmetic all numbers are represented by collections of 1s. Thus 1 represents the ordinary decimal digit 1, 11 represents 2, 11111 represents 5, and 11111111111111111 represents 17. In unary notation it does not matter in what order the digits are read; they are all of equal value. Like a heap of stones, unary digits can all be taken together as a collection: *where* the digit occurs in the string of ones does not matter, since they are all of the same type and each represents the same unit quantity.

How is this coding done in DNA? The most straightforward possibility might seem to be if each base (A, T, G, or C) was used to represent a single digit. But if information were coded in this way, there would be long sequences of As, Ts, and so on, which do not seem to occur in extended repeats.

An alternative possibility might be that the bases A, T, C, and G used in DNA represent four different digits. Thus if A = 0, T = 1, C = 2, and G = 3, then the sequence TAGC would be the number 1032 in the base-four system of arithmetic.[5] Why should such a system not have evolved? The answer, I think, is that this kind of code would be very sensitive to noise introduced by accident or to interference by a hostile agency. Suppose that the initial T is changed by some malignant retrovirus into a G. The number would now be 3032, a grossly larger quantity. Nature is usually very chary of putting all her eggs into one basket, either literally or metaphorically, and is as careful as we are when we transmit information using codes. Usually we introduce a lot of redundancy—extra parts of the message that ensure that we are understood correctly. Sometimes this is done explicitly in the form of check digits; sometimes by repeating part or all of the message in the same or different ways.

A code that would in some ways be more sophisticated and in other ways simpler might use binary arithmetic. This would be simpler because the bases always link together on opposite strands of DNA in the same pairings—A is always paired with T and C is always paired with G. So if an A–T pair was coded as 1 and a C–G pair as 0, the DNA could represent digits in the binary system. For example, the sequence AGGCT would read 10001. This coding also offers some redundancy as a check against corruption: if an A was transposed into a T, the number coded would be the same as before.

Neither of these codes seems very likely. The repeats that do occur are usually of units larger than single bases or doublets, and sometimes very much larger. As we have seen, repeat sequences occur frequently throughout the genome. Such sequences are of varying types and proportions in different species, and the size of the repeated sequence also varies a lot. For example, repeats of the sequences CA and GA are found in many eukaryotic species: TCTCC and GGAAG are found in the chicken, and ACAGGGGTGTGGGG is found in humans near the insulin gene. In view of this, it seems most likely that we should identify a *sequence* with a unary digit: a single repetitive sequence of DNA would then stand for a single unary digit and a string of them would represent a unary number. This is the most straightforward possibility and the one that might be most likely to occur in nature.

It is even possible that the number of bases in the repeat unit may not be a factor of crucial importance. If there is a mechanism in the cell that counts repeat units, then each could be interpreted as a unary digit and the sequence of repeats could represent a number: the greater the number of repeats the larger the number represented. It might not even matter what the unit of repeat was: *any repeated sequence might be counted and registered as a unary digit.* This would provide a very flexible way of representing vital information. Even though there is no compression of information as there is in decimal or binary codes, the enormous length of the nucleic acids means that their capacity for carrying information is also huge. I said at the start of the chapter that a biological computer would be far more powerful than any man-made device. To give an idea of this power, it has been estimated that perhaps 100,000,000 base pairs in mammals have at present no known function in expressing protein. If this redundant DNA is in fact a part of the computing mechanism, then there is tremendous power available. Thus, if in each cell of the organism there are 100,000,000 such locations, with 100,000,000,000 cells in the body of an advanced organism, this represents a possible storage capacity of the order of 10,000,000,000,000,000,000 (or ten quintillion) bits of information—well beyond the most ambitious designs of present-day computers.

THE TURING MACHINE— A UNIVERSAL COMPUTER

Assuming that the organism is capable of holding information that could be handled by a suitable computing mechanism, the next question is: what sort of mechanism would this be? One such mechanism was described by Alan Turing[6] long before the structure of DNA was known, and it has since been recognized as the universal prototype of all computers operating at the nonquantum level.

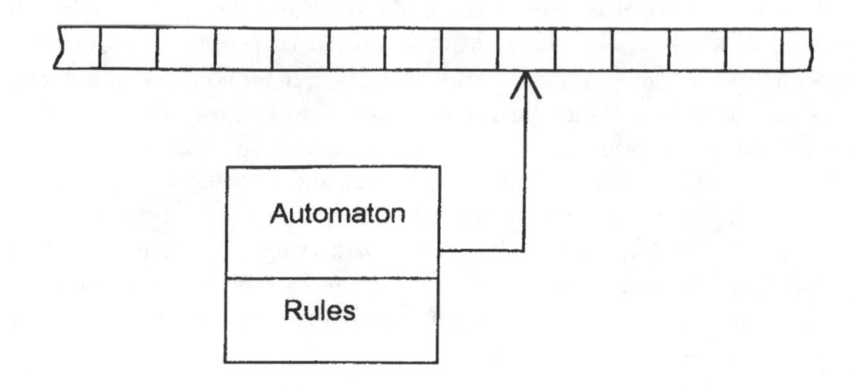

FIGURE 7.1 The elements of a Turing machine.

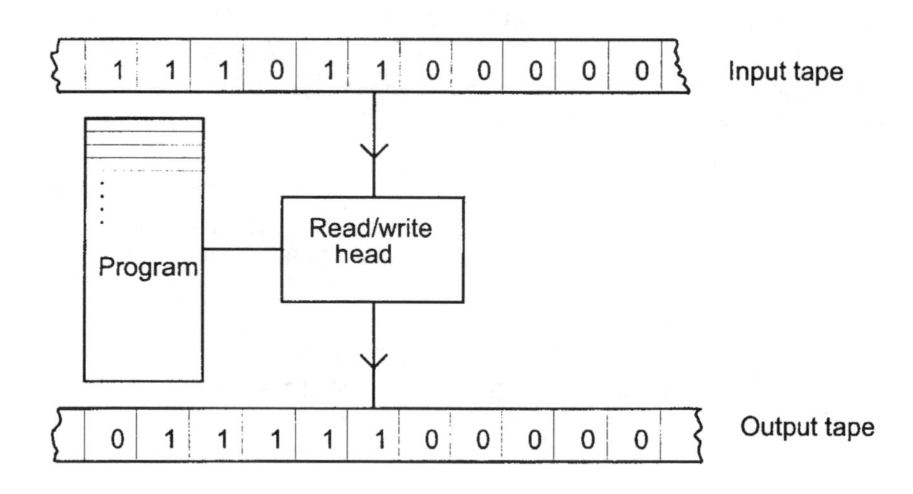

FIGURE 7.2 A simplified diagram of a two-tape Turing machine.

The Turing machine, as it has come to be known, is a notional device, which was never actually built for serious computing purposes, but it describes the operations a computer is capable of performing.

A Turing machine consists of an automaton with a movable tape, originally conceived of as a long strip of paper on which symbols can be written in pencil. The symbols can be either read or erased and rewritten by means of a read/write head. The other principal component is a device that holds the program of the machine's operations. This is a "black box" with the rules for reading, writing, and moving the tape. The rules prescribe an action for each symbol read, depending on the value of that symbol and the state of the machine. One of the rules tells the machine to halt when a desired result has been reached. Figure 7.1 shows the elements of this Turing machine.

In this form of the Turing machine, the tape symbols are written in unary arithmetic, in which a number is represented by a string of ones; for example, 5 is written 11111 and 3 is written 111. In order to separate the numbers from each other an additional character is required, which can be represented by a zero (but which is not a true number zero). So the number 53 would look like 111110111. This system looks superficially like binary arithmetic, which also uses only ones and zeros, but the similarity is deceptive. An important difference is that in binary the symbol 0 is used as a digit; in unary it isn't, it is a separator character or stop marker meaning the end of the number. Instead of a zero, a space could be used as a separator, as in writing ordinary numbers. A further feature is that any number of separator characters is equivalent to one separator, so that the unary numbers 1111100111 or even 111110000000000111 would also represent 53 (this is of potential significance for the biological representation of numeric quantities).

Another version of the Turing machine has two tapes, one that is read and one that is written. The input tape contains the data and the intermediate and final results are written on the output tape. A simplified diagram of this machine is shown in figure 7.2.This version of the Turing machine is just as computationally powerful as the one-tape version.

Subtraction is worth a little attention. Subtracting two numbers is usually done in computers by adding two numbers, one of which is negative.

So 3 – 2 should really be written 3 + (–2), where the number –2 is being added to the number 3. This requires a way of representing a negative number inside the computer and the usual way is to have a register of fixed size and to interpret any number that has a digit in the most significant position of the register (the sign bit) as negative.

In unary arithmetic there are other ways in which we can do subtraction. We can set the number to be subtracted—called the *subtrahend*—beside the number to be subtracted from—the *minuend*—and cut off the number of digits in the subtrahend from the number of digits in the minuend, and it is easy to imagine physical processes that could do this.

	UNARY representation	DECIMAL equivalent
Minuend	1111111	7
Subtrahend	1111...	4
Result	111	3

Multiplication of two numbers in real terms would consist of copying out a number of times a string of the length representing the first number, the number of times being determined by the number of digits in the second.

THE CELLULAR TURING MACHINE

Is there the possibility of something like this occurring in the biological cell? From what has been discussed, it is apparent that many of the elements of a Turing machine are indeed present in the cell nucleus. The most fundamental operations are reading and writing a symbol, which makes possible the copying, replacement, and erasure of old symbols on the tape. These are also basic operations in molecular chemistry. In the Turing machine model presented here, the tapes correspond to nucleic acid strands, the symbol alphabet is the set of bases available (A, G, C, T), the read/write head is an enzyme (a transcriptase), the transcription process is the operation of the computer, and the computation is the sequence of symbols in the transcribed nucleic acid. If the code is regarded as binary, then the alphabet of symbols would contain some redundancy, since four bases are used to represent two symbols. But as we have seen, there is good reason to believe that the coding system for numbers in a cellular Turing machine might be unary, the number of bases involved in coding a digit possibly varying from one situation to another (which might explain some of the redundancy of the genetic code.)

If the operations of transcription were simply those of copying or inverting sequences of bases, it might be hard to see how any computing except that of a trivial kind could be done by means of transcription. However, there are a number of other processes that offer the opportunity for useful computation of a more advanced kind to take place. It is known, for example, that guide

RNA can edit messenger RNA (mRNA) sequences by insertion of bases, including bases from its own tail.[7] Inserted guanosines, giving rise to repeated guanosine sequences in the phenomenon known as polymerase stuttering, are known to occur in certain viral RNAs.[8] Stuttering, or repetition of sequences, is an obvious way in which the multiplication of numbers in unary code could be performed. Similarly, other editing operations such as deletion, insertion, or substitution of components are common in sequential transcription and translation processes.[9] Addition in a unary system is a matter of simple concatenation as discussed above, and testing of magnitude can be accomplished by comparison of base sequences with their maximum permitted length. RNA is edited, copied, and spliced; multiple copies are made (RNA repeats) and strands are cut or ligated. All the necessary biological operations that correspond to the mathematical ones exist in the processes undergone by RNA.

MORPHOGENS, GENES, AND PATTERNS

There are other possibilities, of course. One is that the pattern formation is the result of an analogue rather than a digital process—this was the proposal made by Turing himself. He suggested the existence of chemicals called morphogens, whose concentrations could determine organic shapes and their patterning, such as the stripes of the zebra. Mathematically, the model was a success but, although short-range morphogens have been identified, long-range chemicals that could act as morphogens remain elusive.[10]

The mechanism of action of a putative morphogen is as follows: first, the morphogen or morphogens are diffused from their site of origin in the developing organism. As they diffuse away from this point they become more dilute, either because they are filling an increasing volume or because they are lost by absorption, reaction, or in some other way. Within their sphere of influence the morphogens prescribe a particular mode of growth or development. Where the morphogens meet there may be a reaction between them and then other modes can arise; this is referred to as a reaction–diffusion model.

A classic case is the protein Bicoid, which is supposed to prescribe positional information in the fruit fly *Drosophila melanogaster*. Studies by Driever and Nusslein-Volhard[11] suggest that Bicoid is emitted from a site at one end of the fly embryo and then diffuses throughout the organism (crossing the boundaries between cells is not a problem, because at that early stage of development, the cell nuclei exist in a syncytium—a sort of cytoplasmic soup). In this way, large differences in Bicoid concentrations are set up between one end of the embryo and the other. Bicoid had been thought to give rise to the expression of the *hb* gene, which expresses the protein Hunchback.However, the distribution of Hunchback, intimately linked to the final morphology of the organism, is much more precise than that of Bicoid.[12] The equivalent levels of Bicoid vary between one individual and another by as much as 30 percent of the total embryonic length, while the level of Hunchback, essential for prescribing shape, is precise to about 1 percent. Clearly the information conveyed by Bicoid is not sufficiently accurate for it to act as a morphogen by itself.

More recently it has been suggested that the protein Wingless acts as a morphogen in *Drosophila*. The presence of Wingless in the imaginal disks of these flies is associated with the development of their typical wing pattern. However, for it to determine the shape of the wing in this way, Wingless has first to be expressed, and for this to happen the messenger RNA necessary to produce it must be present. Many developing tissues are composed of epithelia or outer skins that are polarized, so the proteins they contain are located at one end or the other. It turns out that the distribution of Wingless is determined by the location within the cell of its mRNA. If the mRNA is altered, then the protein expressed fails to signal adequately. The information required for signaling is evidently dependent on the mRNA preceding protein expression.[13] This may be the mechanism used in other species also.

Messenger RNA directs the protein to its site of function within the cell. How does it get to that location? After transcription, mRNAs are released and diffuse outward throughout the cell. To reach the site of expression, the mRNA is transported along microtubules,[14] like a packet in a pneumatic tube in an old-fashioned office or department store. This depends on a *motor protein* called dynein, without which the required movement will not occur; dynein-directed transport seems to be a widely used mechanism for the delivery of mRNA. Dynein is the postman of the cell, delivering mRNA packets according to their address labels. This means that positional information is in the mRNA itself. Chen and Schier[15] experimented with the proteins Squint and Cyclops, which are thought to be required for mesoderm formation and patterning in the zebrafish. However, their method of inducing Squint and Cyclops was to inject the mRNA for these proteins, not the proteins themselves, leaving open the possibility that it is the nucleic acid, rather than the protein it expresses, that is the morphogen.

Scar tissue rarely disappears; it is like a kind of synthetic filler material that is used to plug the gap left by a wound. These cells do not know how to develop like the original ones. However, beside the damaged area, the cells know where they are and can gradually close the gap if it is not too large. In *Arabidopsis thaliana,* another much studied species, the protein SHR is essential both for cell division and for the specification of cell fate in the endodermis, a layer inside the developing root. Is SHR a signaling protein, in other words, a morphogen? That depends on how the SHR gets to its site of influence. A study by Nakajima and others[16] showed that SHR moves to an adjacent layer without its mRNA. Plants and animals are different in many ways, but it seems likely that proteins can specify cell fate in adjacent layers, though probably no further away.

The analogy of nucleic acids to the tapes of a Turing machine suggests a very different approach to morphology. A recent review concludes that despite the investigation of several candidate morphogens, none of them fulfill the criteria a classical morphogen would require.[17] A cellular Turing mechanism, if one could be found, would go some way toward clearing up the mystery.

ORGANISMS AS SOLUTIONS TO "HARD" PROBLEMS

What function might such a cellular computing mechanism serve? An obvious answer to this question is that it might provide a means of determining morphology in general and development in particular. Various strands of evidence make this answer convincing. First, for reasons of economy of representation, it is more likely that a computed formula for morphology would have evolved than any other kind of representation. Another argument for computing as a morphogenetic mechanism is the consistency found in biometrics. It has been observed in many different species that there are widespread invariants in biometric measurements. Such consistency can best be accounted for by a computational process that was carried out during development, a process, moreover, that is shared among all the members of a species. Structure in plants is often precise, even when no adaptive advantage can readily be seen. It is plausible that such structural features are the outcome of the complex processes of mathematical transformation going on in the cell.

So it seems likely that DNA contains a Turing machine—a means of computing a function. Each organism embodies a unique function. This function probably determines morphology but may also determine other characteristics through gene switching; either individual genes or combinations of genes may be turned on and off to produce the expression of characteristics. We can see that a computing mechanism as a control device gives great power for expressing almost any characteristic.

What sort of computations is this biological computer performing? Lipton[18] shows how it is possible to construct a computer based on DNA that will solve computationally "hard" problems such as the traveling salesman problem.[19]

Lipton's method uses a number of test tubes, into each of which are put strands of artificially constructed DNA. The strands correspond to a graph representing the problem to be solved; some of the strands represent the vertices and some the edges of the graph, and there are also strands representing the starting point and the desired goal. After a sufficient time, the recombination of these DNA strands will lead to a solution in the form of a new strand of DNA, which can be extracted from one of the test tubes. The important thing is that the time needed to solve the problem by this method is much shorter than could be achieved by any electronic computer known at present, even if a very much faster machine could be built. The reason for this is that the test tubes are all carrying out the process at the same time: in computer terms it is a parallel processing machine (the fact that it is a biological machine does not prevent its being a machine). Because all the operations are done simultaneously, the time needed to solve the problem is the time of the rate-limiting process, that is, the one that yields the correct solution.

This is a theoretical type of computer, and not one that can be constructed at present, but the principle is important. A molecular biological computer can, in theory, perform very hard computations in a reasonable time, mainly because many operations can be performed in parallel. So it is possible for biologically based computers to solve problems that cannot be solved in any way known at present.

PROBLEM SOLVING IN A TEST TUBE

Lipton's "computer" uses a known property of DNA combination, called *annealing*. This is the process that occurs when DNA is heated: it separates into its strands, and when it is cooled annealing takes place. Bases will re-link with one another according to the normal pair-bond rules, forming a new double strand, perhaps with some missing links. So the time taken in Lipton's process is still long, because the process of recombination is essentially an undirected one: we have to rely on massive parallelism in order to produce an interesting or desired result.

Is there an alternative mechanism available that might harness the immense parallelism of even the simplest multicellular organism, but could improve on the random-selective approach of Lipton's biocomputer? The shuffling of nucleic acids that takes place during transcription, translation, and reverse translation may offer an opportunity to do useful computing. Suppose we have two strands of RNA that hold the representation of numbers in unary arithmetic. If we join these two strands, we have one that represents the sum of the two numbers. If we can chop off a length of RNA from one strand corresponding to the length of the other strand, then we have subtracted the two numbers. Multiplication is a little harder: we would have to copy one strand the number of times represented by the other strand. Of course, it is possible to do computing that uses only addition and subtraction, but multiplication shortens things, so we will include it in the basic instruction set of the biocomputer.

Now we have a very powerful computer, essentially identical with the Turing machine, and therefore capable of doing everything that a Turing machine can do: that is, any computation that can be expressed iteratively. It is at work in every cell of the organism, and it is powerful enough to do arithmetic corresponding to the calculation of simple formulas and, by means of iteration, to compute very complex outcomes. Furthermore, it is not one we have to build—nature has already done that for us. This computer was designed when the nucleic acids and their rules of manipulation evolved. Using DNA as their computing formula and RNA as their register, the cells have all the necessary information to perform an iterative calculation based on their position.

MULTIPLICATION—NESTED REPEATS

As mentioned earlier, the patterns of repeats are sometimes more complex than simple repeat sequences. Often these take the form of nested or higher-order repeats, with a repeated sequence that is itself then repeated several times, forming repeats-within-repeats. These could provide a ready basis for representing the multiplication of two numbers: the number of units within the repeat represents one of the numbers and the number of repeats of the unit represents the other number.

The repeats in bovine telomeric and centromeric satellite DNA follow the pattern shown in figure 7.3. Although the sequences are not identical in number, with the numbers 28 and 30 interspersed, the 28 and 30 sequences may be close enough to be considered equivalent. Redundancy is a general feature of genetic coding—it contains unnecessary bits, as noted previously. Often sequences that are roughly the same with one or two variations are equivalent in their effect: a simple example is the fact that the codons ACA, ACG, ACT, and ACC all represent the amino acid threonine. In view of such a redundancy of coding, it is plausible that sequences that are substantially the same—say 90 percent similar or more—might be recognized by cellular mechanisms as being equivalent for counting purposes. This pattern is then suggestive of a number representing the product of a multiplication: there are 4 repeats of about-60-base-pair sequences embedded within a much longer repeating sequence.

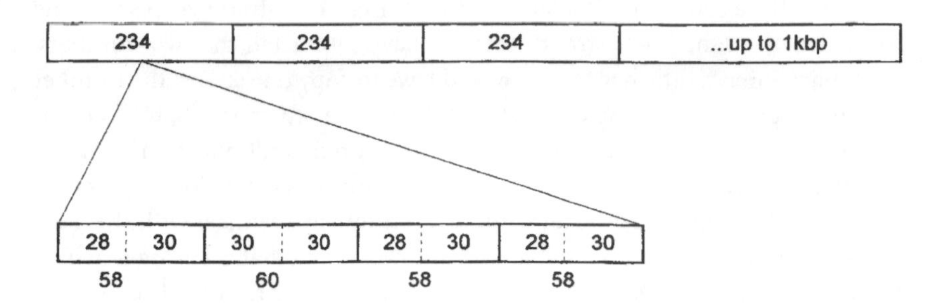

FIGURE 7.3 The repeats in bovine telomeric and centromeric satellite DNA.

The centromeres are active during cell division or mitosis, when two chromosome strands separate to the two parts of the dividing cell. Centromeric DNA consists of long sequences of repeats, making it difficult to sequence. Schueler and others[20] have succeeded in unraveling some of this complexity in the case of the human genome. In the region between expressed genes and the flanking array they found a region of about 450 kilobases consisting of what is called *heterochromatin,* which contains repeating sequences. Each repeat was about 171 bases long and almost the same as the other repeats. In the center of the region the array changed to a higher-order one in which the 171-base repeats themselves were repeated about 12 times, followed by a repeat of the entire sequence. This higher-order sequence can be transplanted into artificial chromosomes, where it will function as a normal centromere during cell division.

THE CELL CYCLE

The development of an organism is not simply growth; it is also a shaping process. The organism reaches its final shape because certain parts grow more than others and some parts die away, like the webbed fingers of embryos. Development is a combination of two processes: somatic cell division, or mitosis, and controlled cell death, or apoptosis. In order to understand morphology, we must first understand mitosis and apoptosis. If the computed morphology hypothesis is correct, then the biocomputer should intervene at certain points in the cell cycle governing these two processes.

There are many critical points in the cycle, but the two that seem most likely[21] to be involved in a computational decision to divide or not divide are the stage after which DNA is replicated (Start) and the later stage at which maturational promotion factor (MPF) becomes active (G2). The timing of MPF activation is known to depend on a number of physiological factors. Presumably a computation is performed during DNA replication after Start, and the result of this is tested at G2 to see whether or not the cell should divide. The microtubule organizing center (MTOC) is also replicated after Start.

In somatic cells, if DNA replication is prevented from occurring, the cell cycle is halted, while in embryonic cells, blocking DNA synthesis does not prevent the cell cycle from continuing. Interestingly, embryonic cells lack a Start phase in their cycle, and mitosis is initiated by inactivation of MPF, followed by DNA and MTOC replication. Perhaps in such cells a later decision is taken to allow the cell to die if it has divided incorrectly, to allow as rapid a decision as possible with a default situation of growth rather than no-growth.

What kind of program is being obeyed by this biocomputer? I have proposed that one function is to make us the shape we are—that is, a dendritic shape. A dendritic or branching structure has, as we have seen, many advantages. Connectedness is guaranteed; distribution (of blood, oxygen, nerve impulses) is achieved in an economical way; wear and tear are minimized; and locomotion (hydrodynamic, aerodynamic, or on level ground) is assisted.

Such a computation of shape by iterated fractals requires two kinds of information. First, the cell must know its position and orientation; the other information required is the formula to be iterated. These two pieces of information will have independent representations within the cell.

When a cell divides, each of its descendants knows its orientation[22] and that it is the result of mitosis. It can therefore know its position relative to the parent cell and, if this information is carried forward in the cell line, relative to a mother cell from which the organ or organism developed. A decision to divide may be taken toward the end of the G2 phase, dependent on certain signals under the control of the p21 protein, itself triggered by the p53 protein, which responds to evidence of DNA damage and signals of hypoxia and oncogene activation. Providing none of these conditions prohibits it, a decision to divide or not to divide may then proceed under computational control.[23]

This iterative computation requires three things: biocomputed formulas yielding a repeatedly iterated complex variable, a criterion value against which the size of this variable may be tested, and the cell's positional information. The question naturally arises: where are the realizations of these located? The formulas and the criterion against which the complex variable is tested are hypothesized to be in DNA, since it surely contains the (normally) well-conserved program for body plan. The variable is probably located in RNA—this acts as the register of the biological computer. The location of the criterion is more obscure: presumably it is in protein and not RNA, as reverse translation from protein to RNA is not known to occur.

As we have seen, cellular positional information can only be known relative to other cells, and must be updated on cell division. It seems reasonable to suppose, therefore, that before a cell divides, it knows its position relative to surrounding cells. This information may be recorded in the histones, which are thought to regulate DNA functioning[24] and are complexed with DNA during the G2 phase. This suggests the following way in which computation may be carried out during the cell cycle.

The size of the variable is the modulus of a complex number representing the cell's fate (not its position, although its initial value depends upon position). If this value becomes too great under iteration, then the cell ought not to survive in that position. This is the value that is recomputed on each cell cycle and is tested against the criterion. The criterion information is used for comparison with the computed variable: if the variable is larger than the criterion, then apoptosis will occur; if the variable is less than the criterion, the cell will maintain itself or divide depending on whether it has neighbors in the direction of intended division.

Any of these factors may be functioning normally or abnormally. They are functioning normally if they are set to the correct values; incorrect values will result in errors with consequent abnormality.

The iterative variable size is correct if it is determined by the iteration of the formula, according to the program in DNA. An error in the iterative variable size can occur as a result of alteration in the program due to some damage. Errors in information representing iterative variable size can lead to one of two possibilities:

(i) If the iterative variable is too large, there will be a premature decision to prevent further cell division and the cell will die inappropriately.
(ii) If the variable is too small, the cell will maintain itself or divide inappropriately.

The criterion is normally set to the value prescribed by a combination of the iterated formula that is held in DNA and by positional information about the cell. If the criterion is altered, an error occurs. An error in criterion information may lead to one of two other possibilities:

(iii) If the criterion is greater than the correct value, the cell will maintain itself or divide inappropriately.
(iv) If the criterion is smaller than the correct value, the cell will cease to divide or will die inappropriately.

Conditions (ii) and (iii) would result in cells surviving inappropriately while conditions (i) and (iv) would result in cells dying when they should be maintained. It is therefore a prediction of the computational morphology hypothesis that variation within individual cells of the respective repeat sequences representing the criterion and the iterative variable size will be related in the above way to the fates of those cells. As we shall see, there is evidence to support this prediction.

HOW CELLS LIVE OR DIE: TELOMERIC REPEATS

A region of repeat sequences that has a very important function is to be found at the telomeres, or ends, of chromosomes. The telomeres terminate in several repeats of a multibase sequence. In humans this is the sequence TTAGGG. These telomeric sequences are of special interest; they seem to control the length of life of the cell. When the cell divides, the number of repeats decreases, and when the repeat number falls below a certain threshold (called the Hayflick limit), the cell ceases dividing. This puts an upper limit on the number of cell divisions that can be performed—about 40 for most types of human cells. The cell then enters a state where it maintains itself but does not divide further. However, if the number of telomere repeats falls to zero, the cell will die. This cell-death or apoptosis is not a disastrous event, but takes place in an orderly preprogrammed fashion, the cell's components being dismantled and used by its neighbors where possible.

In some cells, however, the number of repeats, instead of falling with age in the normal way, can actually increase. The function of increasing the number of telomere repeats is carried out by a substance called telomerase, which adds copies of the repeat sequence to the telomere ends. This happens in some cancer cells: more than half of all types of cancer cells are known to express telomerase. It is tempting to see the telomere length as an overall mechanism that controls cell division, and therefore not only such processes as aging, but possibly also cancer and degenerative diseases. In cancer there is inappropriate cell division, while in degenerative disease there is inappropriate cell death, particularly the death of neurons. Telomerase treatment is now being tried on human volunteers to see if it will prevent aging.[25] Conversely, telomerase inhibitors are being tried as cancer therapy in mouse models.[26]

Therapies for cancer have also been proposed that involve knocking out the telomerase in the cancer cells that inappropriately prolongs their life. One problem in implementing this idea is that it may also doom other cells to apoptosis—brain, reproductive, and neural cells, for example— that need telomerase to maintain them and without which life for the patient would not be sustainable. The answer may lie in targeting just those cells that are inappropriately positioned, and destroying the telomerase only in them. Such cells might be distinguished by the fact that their computing mechanism has gone wrong (due to damage caused by viruses, toxins, radiation, or other trauma), yielding the wrong positional information.

On the degenerative-disease side of the equation, it is interesting that thalidomide, a drug notorious for the morphological defects it induced in fetuses, is now being used as an anticancer drug.[27] Its mode of action, while not yet fully understood, may be due to its power to cause apoptosis, perhaps particularly in preventing angiogenesis, the development of new blood vessels to supply the cancerous growth. Once again, if only those cells that are inappropriately dying by apoptosis could be targeted, they might be saved, or even new cells generated. Again, the targets would be those cells where something had gone wrong with the machinery by which they do their computing.

The importance of positional information to cell survival is clear from a recent study by Weaver and others.[28] The vulnerability of target cells to drugs and other inducers of apoptosis depended on their place in a three-dimensional structure. Tumor cells, lacking such a structure, were more liable to apoptosis as a result of exposure to these agents, while normal cells that had positional cues survived.

REPEAT SEQUENCES IN DISEASE

A class of degenerative diseases known as trinucleotide repeat diseases[29] are produced by an expanded number of repeats of triplets of DNA. In these conditions, which include Huntington's disease, myotonic dystrophy, cerebrobulbar ataxia, and fragile X syndrome, the common factor implicated is an increase in the number of repeats of a group of three nucleotides in the relevant gene. The repeats GAG and GAC are those most usually associated with the diseases, but CGG repeats have been found[30] in fragile X, where protein binding was inhibited by complete methylation of the trinucleotide repeat. This is also a feature of disorders in which the cell decides to die in response to one of a number of signals such as hypoxia, oncogene activation, and DNA damage,[31] all of which are mediated by the regulator p53, a final common pathway for challenges to the viability of the cell.

In the case of Huntington's disease, the trinucleotide repeats are expressed in a protein called huntingtin, which in turn contains expanded repeats of the amino acid glutamine. It is still not clear how huntingtin achieves its effects, since it is anatomically widely distributed, yet it affects the brain specifically. The authors of two recent reports[32,33] conclude that huntingtin probably works by affecting the transcription of certain genes. If there are more than 50 glutamine repeats, the disease will occur; if there are less than 48, it will not. It also seems likely that some degenerative diseases and apoptosis are linked.[34]

ONTOGENY (ALMOST) RECAPITULATES PHYLOGENY

The factors influencing development must be intimately linked to those influencing evolution. The appreciation of the importance of this link has led to the development of the movement called Evolutionary Development, or *evo-devo*. If a computational theory can explain the shapes of existing forms, then this connection suggests that it has also affected the way in which these forms evolved. We now turn to the question: if morphology were determined by computation, what part would this play in evolution?

During the growth period of an organism we assume that its cells are dividing according to the rules already discussed. The placement of new cells will ultimately be determined by the value of a computed number tested against a criterion. The computation will proceed, I have assumed, in the same way for all cells in the cell cluster, the resulting shape of which will therefore be a fractal determined by the mathematical process of iterating this computation. The iteration number currently reached is clearly one of the most important determinants of morphology. At each cell division a comparison must be made between the value produced by the iteration of an expression and a criterion against which the value of this expression is tested to determines whether or not the cell is to grow. This criterion value is an important factor. The other factor that must be considered is the iteration limit. Suppose that there is a maximum number of iterations to be performed. The organism is mature when this number has been reached. During the evolutionary process there may well have been changes in any of these parameters: the formula itself, the criterion value, and the maximum iteration number, or iteration limit. We may assume that in this model evolution is mutation-driven, so that new shapes are constantly being thrown up for selection. As the number of iterations to maturity increased, so would the form

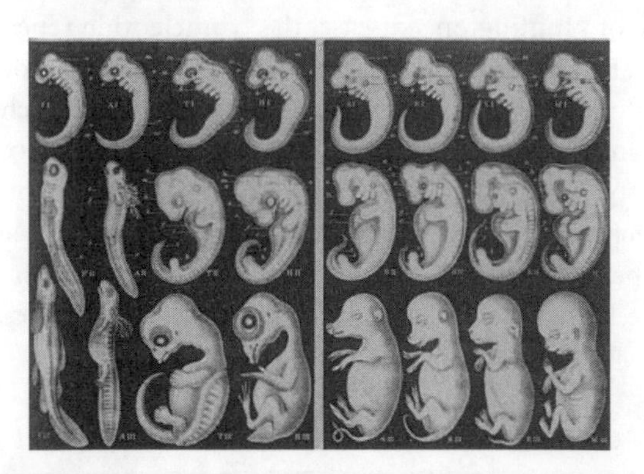

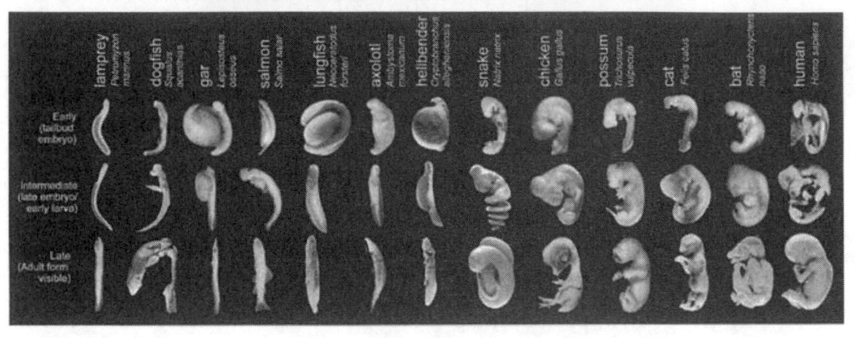

FIGURE 7.4 Haeckel's drawings (above) and actual embryos (below).
Kindly supplied by Professor Michael Richardson.

of the dendritic systems, including the soma, as a greater depth of iteration was used in the determination of morphology.

The doctrine of Ernst Haeckel, that "ontogeny recapitulates phylogeny," that is, that the history of evolution of a species is rehearsed in its developmental stages, was based on the similarity of embryos from widely differing species, which he took to imply that the developing embryo of a higher species passed through these lower stages first. It has been suggested by Michael Richardson that Haeckel exaggerated this degree of embryo resemblance in his drawings.[35] When the drawings are compared to the actual embryo specimens, it is clear that Haeckel indeed highlighted those features that he wished to emphasize. Whether he did so to "fudge" data or from a conviction that there was a common underlying sequence of form is not certain. Figure 7.4 shows the actual embryos and Haeckel's drawings. Clearly there are important differences between the two.

But this divergence from strict recapitulation might make us consider another possibility: that during evolution there will be not only parameter changes that produce small changes in the organism, but also more fundamental mutations that affect the final developed form of the biomorph. These mutations will be those that shape the different species, including their subsequent phylogenetic line. We are used to thinking of species as if they are in one connected tree of descent, and mutation as shifting a species at a node from one branch to another. In actuality *a mutation shifts a species to a new and disconnected path*, so that its ontogenetic sequence will now not quite match its phylogenetic sequence. Such a jump from one lineage to another will result in one species only approximately resembling another during the early stages of growth. The forebear of humans, for example, would not have resembled the exact form suggested by the early state of the human embryo. So perhaps Haeckel's idea was right, even if he exaggerated the likeness of different species in his drawings, which might be better understood as illustrating the idea rather than the species depicted.

FEAR OF SYMMETRY

If the symmetry of the dendritic form is an advantage, then why not have perfect symmetry? Biologically this may be impossible, because the equation that generates the form may not be symmetrical. Or if it is, the part of the form generated by it may display near, but not exact, symmetry. As noted in chapter 2, there must have been a time when the first symmetrical body plan emerged from a hitherto asymmetrical one. The plane-symmetrical plan must have had advantages over and above that of simply having a symmetrical body outline: advantages in having almost complete duplication in the symmetrical disposition of the parts of the body. It is very hard to see how this could have come about as other than a mathematical event.

And here, in this property of symmetry, the same truth applies as to all other characteristics produced during evolution: once it had appeared, the feature of symmetry could be preserved by natural selection, but such a feature would never arise through such a mechanism. If we reject the mathematical explanation for this symmetry, we are driven to an incredible conclusion: that symmetry, sometimes involving three or more axes, arose through chance variation and was then found to be of such adaptive advantage to its possessors that it flourished and spread.

In Plato's *Symposium* an ancient story is put into the mouth of the character of the warrior Alcibaides: men were once whole creatures consisting of two halves, some one-half male and one-half female, but some male and male or female and female, until the gods, jealous of their happiness in this condition, cleaved them in two, and since then, each has gone about searching for its other half. It is interesting that the evolutionary perspective is the reverse of this; that the symmetry creatures do possess may have been the result of two complementary asymmetries joined together.

Brown and Lander reported on the experiments of Yokoyama and others on mouse embryos.[36] An insertional mutation caused the mouse embryo to develop with a body plan that is the mirror image (enantiomorph) of the normal one: the turning process of wild-type embryos coils to the right while *inv/inv* embryos coil to the left. Brown and Lander speculate about the possible mechanism of this. Their explanation is in terms of morphogens, the hypothetical constructs of Turing. If there is a higher morphogen concentration on one side of the embryo, the center of development could be switched from left to right side. The *inv* insertional mutant was, as Brown and Lander point out, produced serendipitously; this is characteristic of contemporary genetic engineering techniques. An alternative explanation in terms of computation would be that the sign of a term in a formula had changed from plus to minus.

Amar Klar of the National Cancer Institute Center for Cancer Research in Frederick, Maryland has suggested[37] a hypothesis that accounts for the occurrence of patterns in plant growth in terms of asymmetric cell division, where one of two neighboring cells is competent to divide while the other needs another cell cycle before it can divide again. This simple rule results in the spiral pattern of growth following the well-known Fibonacci series, for example, in pine-cones or sunflower seeds.

The only plausible way in which such regularities of form can have arisen is through the expression of a process giving rise to mathematical pattern. Such a process, if it generates certain aspects of the form of every individual member of the species, must be the driving force behind species morphology and therefore, presumably, evolutionary change from one species to another. In other words, it is an inescapable conclusion that many features of morphology are determined by mathematical processes and that evolution results from changes in the value of the parameters determining this process, thus giving rise to variation upon which natural selection may act, if these variations are not sufficiently well adapted to the environment. The changes in parameter values will come about by mutation from time to time, producing jumps in morphology and generating wholly new and complex features in one mutational step.

THE ORIGIN OF SPECIES?

We have seen how under the computational hypothesis every cell contains a biological computer that uses the repeat sequences in DNA to turn cell function, including cell division, off and on, and how the shape of the organism could be worked out using iteration in a manner similar to that originally proposed by Alan Turing. Under this hypothesis, new species appear, not as a result of natural selection, but by mutation largely unselected by the environment. Changes as a result of mutations may occur in three principal ways, all of them affecting the cellular Turing machine counters and program values, but not sufficiently far-reaching to destroy the Turing mechanism itself, an event that would result in destruction of the organization of the cells and the eventual death of the organism. They are (1) changes in the iteration number, (2) changes in the criterion value, and (3) changes in the computed formulas. Evolution in the sense usually understood would be mainly a matter of the third kind of change.

What will happen to an individual member of a lineage when a mutation of the third kind occurs? One would expect from the computer simulations that a sudden change would result because the mutation, since it affects a number in a computation by a necessarily discrete amount, will produce a step-like change. There is no allowance in this model for gradual variation of the type postulated by natural selection. There will, in particular, be no intermediates in cases where the selection theory would predict them. This would explain the absence of intermediates in so many instances of supposed evolution by natural selection. The changes may be more or less dramatic or even exotic in form. Such things as new bones, new organs, or altered shapes of existing parts of the body all may be expected to arise suddenly and without preamble in the case of one single offspring, a new species appearing literally overnight.

APPENDIX: THE TURING MACHINE—ADDITION

To illustrate how the Turing machine works, suppose we want to add two numbers, in decimal notation, numbers 2 and 3 (see figure 7.5). These are stored on the input tape with a separator symbol between them. The initial state of the machine is state S1, with the read/write head positioned over the left-most digit of the number 2. The object of the program is to take the two strings of 1s representing the two numbers and rewrite them so that they form

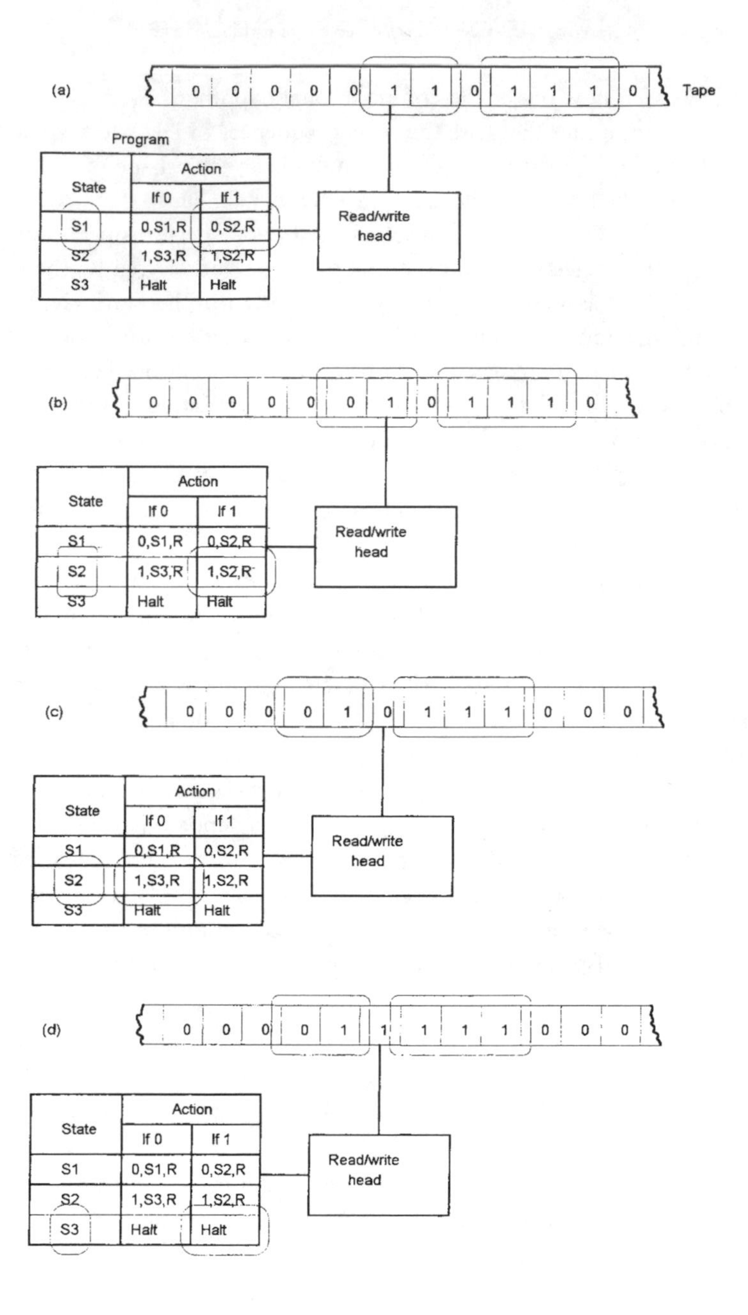

FIGURE 7.5 How the Turing machine adds two numbers.
(For a detailed explanation see the text.)

one continuous string of five 1s. When the program is finished, the machine stops. This string of 1s will then represent the answer.

The steps necessary to do this can be written as a series of instructions to the machine and held in the memory as a program. The program is: the initial state is S1 and the symbol read is 1 (figure 7.5a). The table prescribes the following actions: the new state of the machine becomes S2, the symbol is rewritten as 0, and the head moves one place to the right. On the next cycle the state is now S2 and the symbol read is 1 (figure 7.5b). According to the table, the state of the machine therefore remains S2, the symbol is left as 1, and the head again moves right. The state of the machine is now S2 and the input symbol is 0 (figure 7.5c). The machine therefore enters state S3, the symbol is rewritten as 1, and the head moves right. The machine is now in state S3 and the machine halts regardless of the value of the symbol on the tape. The desired result has now been achieved (figure 7.5d).

The series of operations in this program doesn't just add these particular two numbers together; it is a general program for the addition of any two numbers separated by a blank. In other words, it is a general-purpose program.

We can see that in this way a Turing machine can add any two numbers. It can also multiply any two numbers using a process of repeated addition. Once you have these two processes you can perform any other arithmetic process such as subtraction, division, or exponentiation. This makes the Turing machine very powerful and potentially capable of computing any arithmetic function.[38]

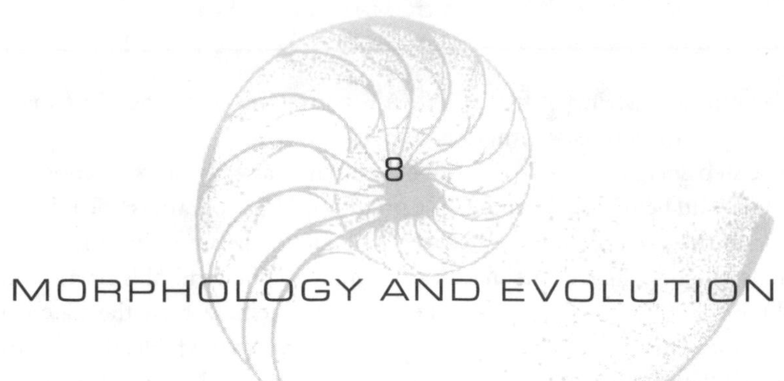

8
MORPHOLOGY AND EVOLUTION

Rejoice in mire. (FRAGMENT 13)

CONSEQUENCES FOR EVOLUTIONARY THEORY

Assuming a mathematical hypothesis of evolution as a starting point for thought and experimentation, what are its consequences?

The hypothesis fits in well with the course of evolution observed in paleontological and other evidence, which, rather than taking place gradually, appears to proceed in steps of a noticeable size—so-called saltatory evolution, or evolution by jumps. This evidence of sudden change supports the evolutionary model that its proponents, Niles Eldredge and the late Stephen J. Gould, call "punctuated equilibrium"; long intervals of stasis interrupted by brief periods of rapid change. Evidence from the paleontological record has been reviewed by Eldredge and Gould[1] and is convincing.

However, the presence in the record of evidence for punctuated equilibrium does not necessarily imply that evolutionary steps, when they did occur, were sudden and large. It could be that, on an evolutionary time scale, gradual change over thousands or even hundreds of thousands of years appears in the paleontological record as a sudden jump and this is Eldredge and Gould's own interpretation. Or it might be that the changes are indeed sudden, but are nevertheless brought about by "gradualist" Darwinian mechanisms.

Various alternatives to saltatory change in keeping with the original Darwinian view have been suggested as explanations for such an evolutionary picture. One is that a number of genetic adaptations accumulate but are not expressed for a considerable period, and then suddenly show themselves, resulting in an apparently sudden leap forward. At first it seems hard to understand how this can happen. On the face of it, an adaptation must express itself; otherwise it cannot be advantageous. So quite how this mechanism could operate is not at first clear. But it has been suggested that epigenetic (between genes) interaction comes into play, so that a number of collectively unexpressed variations might still affect a separate, expressed feature of the organism in such a way as to perpetuate themselves until the accumulated changes can be expressed together in some new feature that then emerges as an apparently sudden change.

PUNCTUATED EVOLUTION?

So far these alternative "saving" mechanisms for a gradualist approach remain theories. However, there is now some experimental evidence accumulating about how evolution actually does occur, if not on the macro, at least on the micro scale, and it favors the saltatory rather than the gradual interpretation. In 1996 Elena and co-workers[2] demonstrated what they call *punctuated evolution* in the bacterium *Escherichia coli* (*E. coli*) by means of a simple yet essential experiment.

A population of unicellular bacteria was founded from a single cell and allowed to develop to a total of about 500 million cells. Each day, a sample of this population was transferred to a new environment. The fate of the cells was tracked for a total of 10,000 generations. Mutations occurred naturally and selective pressure was applied by keeping the environment low in glucose, which is a food source for the bacteria. This had the effect of increasing the mean cell diameter by about 50 percent in the course of the experiment, which lasted for 1,500 days (or more than 4 years) in total. The adaptive significance of the larger cell size is not known, but the larger cells were fitter in the Darwinian sense (tested by being compared with their ancestral cells for growth rate). Since no genetic interchange between individuals was possible (*E. coli* is asexual), any changes in the characteristics of the population must have been due to mutations alone. Mutations occurred at the rate of about a million a day, but most mutations were unfavorable and died out quickly. Seen on a long time scale, the change in cell size was apparently gradual, with size increasing in a series of smooth segments and finally stabilizing (see figure 8.1). The best-fit curve to these data is a hyperbola.

The striking finding was that on a finer time scale these stages of apparently continuous change were not continuous at all but were composed of steps. Figure 8.2 shows the first 2,000 generations of cells on a shorter time scale; the

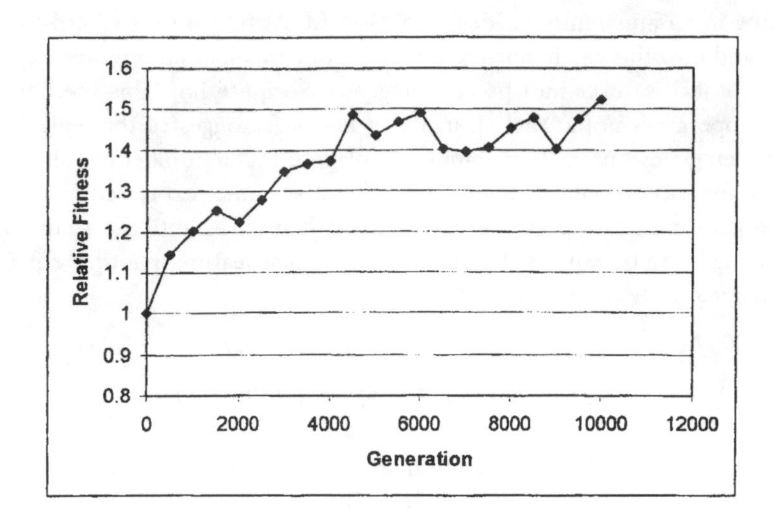

FIGURE 8.1 Continuous cell-size change.
[Recalculated from data used in R. E. Lenski and M. Travisano, "Dynamics of Adaptation and Diversification— A 10,000-Generation Experiment With Bacterial Populations," *Proceedings of the National Academy of Sciences of the United States of America* 91 (1994): 6808–6814, figure 5.]

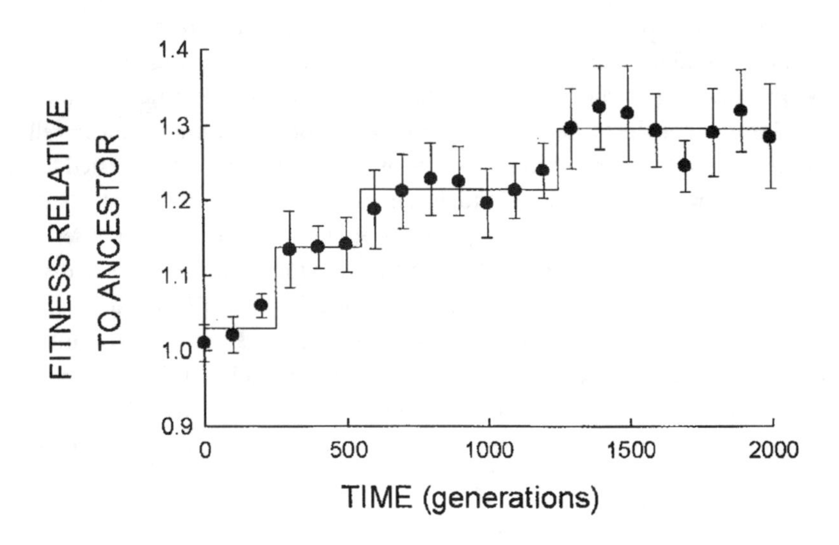

FIGURE 8.2 Discontinuous cell-size change.
By kind permission of Professor Richard Lenski.

size changes took place as a series of steps and not continuously. So, if we can identify evolution with the change in cell size, then evolution occurred not gradually but in steps, along the lines the punctuated equilibrium hypothesis would suggest.

In a previous experiment in 1994 Lenski and Travisano[3] had followed the evolutionary changes in several populations of *E. coli*. Under selective pressure, the bacteria evolved rapidly for the first 2,000 generations or so but then became static and remained so for the last 5,000 generations. The interesting point here is that, despite the fact that the different populations were initially genetically identical and had evolved in apparently identical environments, they diverged significantly from one another in both morphology and measured fitness, coming to rest on different fitness peaks (as they are sometimes called).

Three things are important about these studies, which show unambiguously the way in which such short-term evolution happens. First, evolution went on for a while and then stopped; there was a limit to the change, suggesting that either the selective pressure had been fully adapted to or no further change was possible. (It would be very interesting to distinguish these two possibilities.) Second, the course of evolution in the 1996 experiment (see figure 8.1) showed at least one backward step, lasting for about 500 generations, which is quite a long time. Third, the question of how the changes penetrated the population is left in doubt. One favorable mutation could furnish a new population of 500 million in 30 generations. Since a cell division of *E. coli* takes place about every 3.6 hours, this means that the entire population of 500 million cells could have been replaced by the descendants of a single individual cell in a mere 4.3 days! So each of the changes in the population could have been due to a single mutation in one cell rather than a number of simultaneous mutations occurring in parallel.

Although these experiments were done in terms of micro rather than macro evolution, they show something about the origin and fate of adaptations. The authors comment that they "observed several hallmarks of macro-evolutionary dynamics, including periods of rapid evolution and stasis" and they fit "[an] interpretation, in which chance events (mutation and drift) play an important role in adaptive evolution, as do the complex genetic interactions that underlie the structure of organisms," concluding "for now the generality of our results remains an open question: one might well wonder what outcomes would be observed with a sexual organism."

In their book *Origins Reconsidered*, Leakey and Lewin[4] discuss the speciation event that led to modern man. They distinguish two positions, one that favors the idea of the emergence of humans in many different parts of the earth, and another, which they call the African Eden hypothesis, that suggests one place of origin, followed by an outward radiation in many directions from this site. Their comment on the second viewpoint is that it requires the disappearance of the entire precursor species at the same time as the emergence of the new one, presumably because of the Darwinian concept of natural selection by reproductive success. It is possible to see an alternative, however: two coincident mutations that led to a breeding pair, the new population living side by side with the old one, almost equally "fit" at first, but with the new gradually gaining an advantage. The Neanderthals lived alongside us for many thousands of years. It is likely that, whatever our precursor species was, it did not disappear overnight.

SELECTIONIST EXPLANATIONS OF BEAUTY

If new features of an evolving organism are produced suddenly, then their presence may be due to the process that produced them, rather than to any particular selective advantage. All that is required is that they emerge and that they be neutral or at least not grossly disadvantageous. On the gradualist hypothesis, however, they can only have emerged as a result of continuous selective advantage.

It is possible to attribute the existence of any feature to selective advantage if that is the idea you start with. Starting with the hypothesis that all variation between organisms is due to the effects of selection, it is easy to collect evidence for that hypothesis. The more such evidence is collected, the more firmly the hypothesis becomes embedded in the scientific psyche. The thought process goes something like this: "Any feature must be there because it confers a selective advantage: consider feature x; its possible function is Y. This accounts for the existence of the feature; it also provides further support for the hypothesis." A vicious circle like this can continue indefinitely. In the case of esthetic appeal of the sexes, the selectionist argument is that men prefer women who show signs of being able to bear children successfully. For example, facial features that may indicate this will be preferred to others, and this gives rise to the perception of beauty.

Men are repeatedly attracted to the same type of women and vice versa. When people get married for a second time it is frequently to a similar type of person to their first spouse. People with a preference seem likely to go on displaying that preference. Why should that be, if biological advantage was all that counted? It could still be argued that this occurs for reasons of selective advantage. It might be that the shape of the face, for example, indicates that the person is healthy, or fecund, or likely to have long-lasting reproductive capacity. But if so, why do we not all prefer the same type of mate? Sexual preferences are very varied; one person's preference is not the same as another's. How could this be if all that mattered was the same set of facial or bodily visual signals? If a big nose, say, is of adaptive advantage, is this not equally important to all the potential mates of the person having that feature?

It could again be argued that what is of advantage to one person may not be to another. A big nose may confer the advantage of rapid air intake, but it will not filter the air as thoroughly for particles. Perhaps what one person needs in a mate is a big nose, because their family has been plagued with respiratory difficulties, while another needs a small-nosed mate because their family has suffered from bacterial illness brought about by insufficient filtering of the air inhaled. Such things are difficult to quantify and there are many alternative explanations that suggest themselves. An additional idea is that features that have no obvious adaptive significance may be related to health and thus may be certificates or advertisements for health. For example, large eyes and small noses are found attractive in women by many men. It is not obvious why either should be adaptive, but they could indicate a general state of health. This is the approach of the so-called Evolutionary Psychology movement. Once you have begun to think of explanations of this type, it is not difficult, with ingenuity, to continue indefinitely, but this very easily becomes a self-serving bias of thought. A recent paper on evolutionary psychology asks baldly "Why are human brains designed by past Darwinian selection to infer that attractive people are more valuable social resources than less attractive people?" With the questions (and answers) so blatantly begged, there is only one answer to that.

WHAT IS BEAUTY?

An alternative view is that beauty is the result of a correspondence between the contemplator and the object of contemplation, more specifically between the nervous system of the perceiver and the morphology of the perceived. What kind of correspondence could this be and is it likely that the perception of beauty could have resulted from it?

We know that the nervous system, including the perceptual system, has fractal characteristics. We also know that organic forms are fractal, both as a whole and in their parts. Das and Gilbert[5] have shown that areas in the primary visual cortex are organized so that they respond best to rapid alterations in the orientation of a stimulus, while specific points called pinwheels are responsive to many orientations at the same time. This is in contrast to the traditional view of visual receptive fields, which are often conceptualized as linear in form, having been shown in classic experiments to respond to linear stimuli such as edges and bars. Clearly these receptive fields are also well adapted to a primary response to fractal objects, which are often characterized by crinkly, uneven boundaries.

In one study[6] of the sexual preferences of men, three female stereotypes were created by averaging many faces. One stereotype was formed by averaging all faces in the sample of stimuli used; the second was the average of the used faces that Western men found most attractive; and the third was a face formed by taking the differences between the first and the second and exaggerating them. The third face, although corresponding to no known group of people, was the one most preferred. The faces used as stimuli were both Caucasian and Japanese and the finding held for both British Caucasian and Japanese men alike, indicating a general male sense of beauty, while not explaining its significance. The authors comment that attractiveness "may signal sexual maturity" (p. 241); however, female subjects showed the same preferences as males. Among the preferred features was high cheek-bones: it does not at first appear obvious what the survival value of high cheek-bones might be. A more recent study of female attractiveness showed that, while averaged faces are found attractive, the most attractive faces are by no means average.[7] There is also a preference by infants for less average and more asymmetrical faces.[8]

THE EYE OF THE BEHOLDER

Instead of the selectionist hypothesis, let us consider another idea: that the preferred features are those that are the result of a neutral evolutionary change. A mutation affecting the facial features will propagate if it finds a suitable breeding partner. The offspring of the breeding pair will have this trait and it will spread through the population according to its genetic dominance and the reproductive rates of the carriers. Those carrying the mutation will also have their perceptual system shaped by it, in such a way that they prefer partners also carrying the mutation and displaying the trait.

Esthetic responses are not limited to one sense or modality. We take pleasure in visual arts, and this is associated in most people's minds with esthetics. But the tactile sense awakened by sculpture, the kinesthetic pleasures of watching ballet, though they are visually mediated, are sensations of esthetic enjoyment felt in other modalities. Again, the world of music, although in opera and pop music there is an inescapable and strongly visual element, is almost entirely nonvisual.

Marmor and Zaback[9] carried out experiments showing that blind people could form mental images in much the same way as the normally sighted. They had people rest their hands on wooden shapes that were placed in different relative orientations to one another and asked them whether the shapes were the same or different. In half the cases the shapes were not the same but were mirror images of one another. It was found that the blind subjects took longer to respond if there was a larger difference in orientation between the two shapes, suggesting that they were mentally rotating the shapes in order to compare them, a finding that also applies to people making their judgments on the basis of vision. The blind rotated their mental images more quickly than sighted people. It seems that imagery can be induced via more than one sense, leading to a common fund of experience.

Nor is esthetic perception limited to humankind: there is no doubt that some animals also have an esthetic sense. The bower bird creates an elaborate nest as part of its mating ritual. The ornate patterns of birds and fish serve as attractants for mates. Is the behavior of approaching such patterns accompanied by sensations that, if they were expressed verbally by humans, would without reservation be termed aesthetic? Darwin thought so: "When we behold a male bird elaborately displaying his graceful plumes or splendid colors . . . it is impossible to doubt that [the female] admires the beauty of her male partner."[10]

Enquist and Arak[11] modeled the evolution of characteristics such as the flamboyant tails of peacocks using a neural network and found dramatic changes in male tail length associated with female preferences. A 6×6 "retina" of cells with a 10-cell hidden layer was connected to a single output cell. This network was trained to discriminate long-tailed from short-tailed crosses (birds). The network, when subsequently shown test stimuli, gave an even greater response to longer-tailed stimuli, among others, than it did to those on which it had been trained. The authors call this "an inevitable bias in the response of signal recognition processes." The neural network was allowed to evolve (by changing its weights) in accordance with the assumptions that courtship would follow a strong recognition response and that breeding would lead to changed tail lengths.

While Enquist and Arak interpreted their results in the light of neo-Darwinian theory, Heinemann[12] pointed out that the Enquist and Arak findings fit in much better with the views of William Bateson.[13] Bateson believed that features such as these could appear in a single generation and that they were driven by internal mutation mechanisms, rather than externally applied pressure. This model is closer than the Darwinian model to the behavior of Enquist and Arak's neural network. Bateson's views also fit in well with punctuated equilibrium views of evolution.

A general hypothesis may be proposed here: that the perceptual response is related to the structure of the system that does the perceiving. We now have machines, based on neural network models that can "perceive" in the sense of recognizing shapes, but so far there has been no hint of a machine appreciating the beauty of what it sees. Is this because machines are incapable of aesthetics, or is it more likely that, lacking our structure and perceiving shape in a different way, these machines do not have the built-in capacity to perceive those relationships we find beautiful? And if a machine were made that had the necessary built-in structure, would it not also have an accompanying esthetic capacity?

The shapes and proportions that we consider beautiful are of interest because they are linked to the structure of the nervous system and to the way in which that system functions. This appreciation of beauty can be of the most direct nature or it can be quite abstract. It is not only the proportions of shapes that we admire, but also the patterning of melody and counterpoint, and the elegance of a mathematical proof. We speak of the beauty of a well-played game of chess or a well-argued case in law. It seems that beauty resides in the relations between the well-arranged parts of a whole and that we are capable of abstracting this quality and enjoying it from many different sources, just as blind people can abstract and use their imagery from tactile sources as effectively as they could have done from visual inputs.

MATHEMATICAL BEAUTY

There have been many attempts to relate beauty to mathematics.[14] The Greeks believed that beauty in the proportions of visually perceived objects was related to the golden mean. As we saw in chapter 1, the golden mean is based on a line divided so that the ratio of the smaller part to the larger is the same as the ratio of the larger part to the whole. In number theory it has been found that the golden mean is the most irrational number there is—it is the number that is furthest away from any number that can be expressed as a ratio of two integers. When doing experiments in chaos, the golden mean is sometimes used for this reason to drive the inputs to a circuit whose output is then guaranteed to be chaotic. If the phases of the inputs were related by a simple ratio, then the outputs would be also liable to lock into a ratio, but the choice of an irrational number prevents this from happening. If the golden ratio is chosen, the phase of the inputs is as widely separated as possible.

The structure of the nervous system is fractal, the outcome of an iterative process in the development of the nerve fibers. Our perception of beauty is al-y will attempt

inian survival
n goes some-
kely to find a
d on more of-
lacking it. But
eased by some-

orphology. Ac-
ned by mathe-
f the organism.
e prescribed by
is the one per-
perceptual sys-
actual shape of
effects of eating
according to the
on between ideal
o sociobiological

e? Because she ex-
m legs, waist, and
same preferences.
en do? Some men
l, ideal, and most
n overweight, wo-
ncy between their

ETRY

home, largely for
following example,

scription of a pair
s is a fairly unex-
he majority of the

Is it not more likely that what we call
e fractal structure of our nervous sys-
in the survival value, of the object? We
we do not in some way resemble. It
ye and in the brain of the beholder.
admiration of features of the human
stance—usually passes without com-
le in such a predilection. Often, too,
as having an obviously reproductive
1, since people are often attracted to
ave been suggesting is correct, then
orphs of which they are made up is
about their morphology, but also
ms.
version of the Bible for a classic ex-

anteth not liquor: thy belly is
s.
hat are twins.[15]

rs of waters, washed by milk

owers; his lips are like lilies,

s belly is as bright ivory

kets of fine gold: his coun-
rs.[16]

MODELS

he exercises her shape, care-
cent under the nutritionally
t of men, who usually prefer
to be. It is for the benefit of
other women who will em-

ulate her appearance. If these women are not naturally slim, the
to become slimmer, sometimes at risk to their health.

One way of explaining this behavior is to postulate a Darw
mechanism producing the slimming behavior. The explanati
thing like this: women who conform to the ideal are more l
mate, and a genetic tendency toward this behavior will be pass
ten by women exhibiting this type of behavior than by women
this would be at odds with the fact that men would be better pl
one fatter.

An alternative explanation would be in terms of computed m
cording to this notion, the shapes of living things are determ
matical computations carried out by nucleic acids in the cells
In the case of women, the "ideal" shape—their ideal—is the o
the formula carried in their genes. The "most attractive" shap
ceived by men as such because it is embodied in their (male)
tem. The shape of a person can be modified by eating; so th
the woman is the outcome of the ideal (innate) shape and the
habits. In order to maintain the ideal, she will often diet. Thus
theory of computed morphology, there is no necessary relati
size and most attractive size, an inexplicable fact according t
explanations.

Why should the fashion model diet to a less than healthy siz
aggerates and overemphasizes the signs of the ideal, such as sl
smaller bust, in order to sell clothes to other women with the

Why do men not try to become thinner as often as wom
do become anorexic, but in men the three body sizes—actu
attractive—are probably closer, because men are not so ofte
men's weight problems resulting from the greater discrepa
actual and ideal body sizes.

SYMMETRY VERSUS ASYMME

Nowadays descriptions of female beauty usually find thei
want of anywhere else, in the literature of erotica. Take the
which is fairly typical:

> *For as long as I can remember I have found the sight or de*
> *of pretty legs of a young woman an object of delight. Th*
> *ceptional declaration, as it would appear to be shared by*

> *male half of the population, as witness the abundance of magazines, novels, advertising, books on body-care, fashion styles etc., which portray them. By the same token , the majority of the female half of the population would appear to recognize this male interest, and go to some considerable lengths to address this in the way they generally dress and present themselves. I would go further, and admit that my delight can be enhanced by the way the legs are displayed. There is a whole repertory of high-heel shoes, boots, stockings, suspenders, garters, laces, straps—in a vast range of colors, textures and shapes—which would suggest that this interest is unremarkable. Maybe it reaches the proportion of fetishism, but where is the dividing line between this and enthusiastic interest? The interest is not so obsessive as to constitute the only aspect of attraction to a woman. Personality, character, and intelligence count far more, even if the initial attraction may have been a pair of shapely legs.*

But, very interestingly, the writer goes on to comment on his liking for monopedes—women with one leg missing—and attempts to relate it to abstract aesthetic rules:

> *But though symmetry and completeness have their place, what of asymmetry and isolation of the part from the whole? Is a leg in isolation any more strange a concept than a disembodied hand, ear, torso, lips? Isn't there more interest when perfect symmetry is replaced by more complex form, where contrasts are made, where distortions occur?*[17]

It may be that there is a wider liking for monopedes than is usually supposed. This may be yet another area of human sexuality that has been taboo and will, when explored and revealed, be seen to be much more widespread than was previously thought. This was recently confirmed by the emergence of a group of amputation wannabes—people who wish for the amputation of a limb, typically of the leg above the knee. This has a strong sexual component and its onset is sometimes associated with powerful sexual attraction to such a maimed person during early adolescence. It could be argued that this is the cause of the preference: the occasion of exposure to such an unusual erotic stimulus at a critical period determining the sexual partialism. But it is doubtful whether such an event wholly determines the preference; it may more likely be the first sign of its impact as the genetic switch is thrown. Thereafter, a strong identification with the amputated condition emerges and becomes a lifelong obsession, to the point where sometimes a carefully planned act of self-mutilation is carried out. The satisfaction and relief—fulfillment is not too strong a way of putting it—that follows eventual loss of the limb is indubitable and profound.

THE BOY WHO WANTED TO LOSE A LEG

"Carl" is a monopede who had been obsessed with amputation since meeting, at the age of twelve, an attractive boy of the same age with only one leg. He longed to lose the same leg as the boy, but eventually lost the other one as a result of a railway accident of his own making—he lay with his leg across the track in the path of an oncoming train. At first his greatest fear was that the doctors might be able to save the partially severed leg, as they assured him they might—but he managed to maneuver them into amputation. He describes his profound psychological satisfaction at having only one leg. He says he finds some difficulty in managing, but it is well worth it. Carl reported some phantom limb discomfort but it was bearable, he said.[18]

Herman Melville's *Moby Dick* is a book about the loss of a limb and the pain often felt in a missing limb. In the case of voluntary amputees, there is some doubt whether they feel phantom pain or not. In normal phantom limb conditions, the pain may result from the mismatch between the body shape and the body image we hold stored in our brain. In the case of amputation wannabes however, there is a strong internal image of a limb asymmetry, and this will match poorly with the state of the body at birth, but much better with the amputee state. The discomfort that still remains after amputation may be due to learned body image effects clashing with the new and desirable body geometry.

But how could we perceive, much less like, admire, or desire, something that was not somehow already native to our perception? The study of visual illusions repeatedly shows that we cannot accurately perceive objects that are alien to our perceptual system, and this is to be expected because perception depends on the recognition of features. There is a hierarchy of such features and their corresponding recognition units, starting with simple shapes like bars and edges, which can be recognized by cells as straightforward as retinal ganglia, and ranging up the scale of complexity to (it has been suggested[19]) grandmother cells, which respond only when you see your grandmother. Those objects that can be taken in and understood are those of which both nature and the self are composed. The grandmother cell is, so far, a pious hope rather than an actuality. But it is certainly the case that for a wide class of objects there is a mapping between the geometry of the nervous system and the geometry of the object perceived, and it seems reasonable to expect this principle to become more rather than less extended as physiological knowledge improves. One might almost say that it is impossible for the system to perceive anything that does not have the same form as itself.

SEX AND ESTHETICS

But the most significant point is that here is an apperception of a deep truth—that tastes in sexual matters are indeed related to fundamental esthetic questions. I will assert more than this—that they are an integral part of the functioning of our central nervous systems and therefore intimately bound up with cognition, perception, emotion, and motivation, not just tacked on to them as extras. We could go even further and suggest that sexuality is bound up with the structure of not just our nervous systems and our bodies but with the structure of the world as a whole. If the shape of the breast has emerged as a result of the iteration of a formula, the possibility of that form is inherent in any situation in which that formula applies. We name landscape features after such forms: the breast of a hill, the leg of a journey. Even geographical features are often seen in this way: the horn of Africa, the foot of Italy kicking Sicily. And this is not to be greatly wondered at; landscapes both large and small are shaped by physical forces that embody iterative dynamic procedures. Pictures in the fire, indeed! But why are there pictures in the fire? Is not flame a quasi-organic process—yet another zoic system?

THE PROBLEM OF "PARTIAL ADAPTATION"

Another bonus in the mathematical account of evolution is that it explains the old riddle about how half an eye—that is, a partially developed organ—could be an adaptive feature and thus favor the individual with half an eye in the reproductive stakes. The answer is that it can't; but a whole new organ such as an eye can favor the individual; and this is what the individual gets as a result of the mutation and the new computed shape that it initiates. The situation is even that the adaptation is advantageous but the half-way phase might be lethal!

Thus we can make a mental model of evolution taking place in a phase space such as we discussed in chapter 4, a mathematical vector space, in reality having many dimensions, in which are located the states of particular evolved organisms. The dimensions of this phase space would be the spatial dimensions in which possible mutations of the organisms take place, plus time. In this vec-

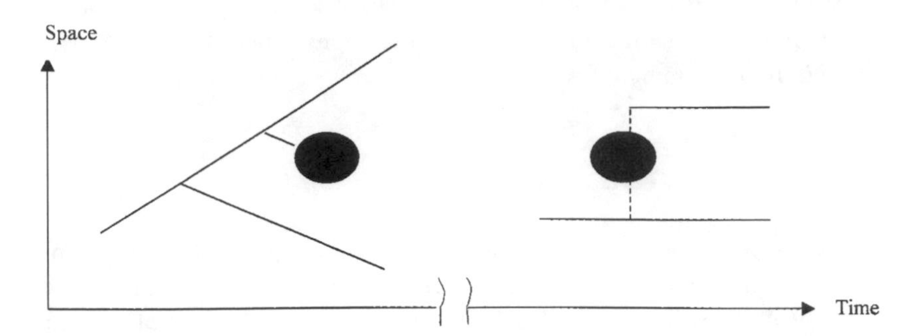

FIGURE 8.3 Evolution taking place in vector space: gradual evolution (left) and saltatory evolution (right).

tor space the individuals of a lineage trace out lines as they evolve. Figure 8.3 shows this space diagrammatically. For simplicity it is reduced to two dimensions, the vertical dimension representing space and the horizontal dimension representing time. Points in this space represent individual organisms at an instant of time—probably rather a long instant.

A mutation takes the lineage of two individuals from one point to another in this space, representing a change in morphology over a certain time and depending on factors like the reproductive rate. Hidden in this space, like mines beneath the sand in a desert minefield, are *lethal zones* of maladaptation, which would kill off an individual having a particular morphology at a particular time. These are sketched in as the hatched areas. Their different effects in the two models of evolution are apparent. If a lineage ends up in one of these, it terminates. However, it can pass right across one of these zones and end up in the safe areas again; this may be exactly what happens in certain types of evolution. Thus half an eye may be disadvantageous or even lethal while a whole eye is adaptive.

What sort of evolutionary changes might take place as a result of changes in the computing mechanism of the cell? In order to give a proper answer to this question one would have to look in more detail at the nature of the computing mechanism itself, but the overall picture is clear. Obviously, some of the results of mutation can be so drastic that the computer ceases to function and the cell dies. But short of this, there will be mutations affecting each of the parts of the computer. The reader will recall from the discussion in chapter 7 that a computer is composed of three main parts: a program of instructions; a central processor that will carry out those instructions; and data including constants and variables for use by the program. Two of these main divisions of the computer—the program and the data—will probably be embodied in the cell as DNA sequences. The central processor will be the chemical processes of copying and concatenating nucleotide chains, which take place under the control of the program.

What would have been the story of human beginnings, if this theory is correct? Perhaps it might have read something like the passage with which this book opened.

The Ascent of Man

About a hundred thousand years ago, perhaps as many as a hundred and fifty thousand or as little as fifty thousand, the first man was born. In the eyes of his parents he must have been an ugly baby, an outcast from the brood and from the tribe from the moment of his birth. He was lucky to survive, for monstrous births were not generally suffered to live in that time and many more like him would have perished, and would in other times and places perish, by accident or by design, and without further consequence. He was different from his tribe. His features to them seemed curiously unnatural: angular, sharp, and distorted. His gracile frame seemed spindly and ill-adapted to survive in the rigorous climate, which could vary every few years, alternating between humid-hot and freezing-dry conditions, interspersed with a temperate mediocrity hardly more favorable, since it ill-suited the growth of plants on which the tribes depended. To the society of the race he was a useless individual—an idle jack—spending much time in seemingly depressed, introverted contemplation, staring into space, fiddling with pieces of stone and slate, or making marks in the earth with a stick. Perhaps his social experience was the origin of the story of the Ugly Duckling. The first Outsider, he must have been familiar with loneliness in a time when loneliness was a difficult condition to achieve and an undesirable one.

Let us call him Adam.

Adam sat for long periods between gathering plants or catching animals, when he was not shivering too much to think or too exhausted by the sun's heat. And in those periods when he was able to reflect, he contemplated the great mystery that was his existence in the world. He could speak in a way—the gift of tongues-that the tribe could not, but he had no one to speak to. He made peculiar guttural and explosive sounds and he seemed to gesture, in apparent madness, at the beasts and plants and rocks, and even at the sky, as he made these strange noises. This was the birth of language, and he was naming the things around him.

Disadvantages Really Advantages

The awkwardness of appearance and different gait of Adam that distinguished him from his parents, although they appeared to be handicaps, were either advantages or reflected some advantage. His thin gracile frame and the development of his feet allowed him to run rapidly, outdistancing his fellows and rivaling the beasts. His fingers, too, allowed a new and easier grasping and manipulation of objects, whether weapons or tools. But the supreme advantages were unseen. Within the odd-shaped skull the brain had developed a further layer of dendritic elaboration that would in time permit new kinds of cognitive skills—notably the use of spoken language and the ability to measure and compute. What could be thought more easily could also be expressed more easily: the artistic impulses that already existed in almost equal measure in Neanderthals now found an outlet in the ability of fingers able to manipulate a brush. The religious sentiments felt by the near-men could now find their expression in prayer. But, above all, the new-found ability to count and group the objects around him gave Adam the cognitive tool with which he and his descendants would master the entire planet. Adam was the first man but also the first mathematician, the precursor of Plato, Leibniz, Einstein, and Schroedinger and in no significant detail different from or inferior to them. With Adam's birth a new world was born also.

He would have died without issue, for he was probably incapable of mating with his own tribe, even if he had been wanted by them (and perhaps he did not regret this), had it not been for the most significant incident in his life and in ours too, for him and for us the most blessed and auspicious of events that could have occurred, giving as it did the gifts of life, love, and hope to the generations to come. He met Eve.

The character of Eve was closer to Adam's than to the tribe, though there must have been differences as well. Adam and Eve were close enough in form to mate and produce viable offspring who retained the distinctive nature of their parents rather than reverting to that of the race. Their meeting and mating may not have been a romantic affair. Far from being another person like him, perhaps she was not very similar to Adam: another misfit but of a different kind, also a mutant and an outcast. Thought unfit for childbearing, she was to become the mother of us all. Rather than the soul mate he sought in his imagination, she was perhaps the only one who would consort with him, or he with her.

The foregoing paragraphs may seem closer to fairy tale than to truth,[20] but they contain a few fundamental points that they express vividly, albeit rather crudely:

1. The human species probably began with a single breeding pair.
2. Both individuals of that breeding pair were the mutant descendants of the same precursor species (identified here as Neanderthal, although it may have been another).
3. Both individuals would have died without issue or their offspring would have been infertile had they not mated with one another.
4. Humans displaced the precursor species largely as a result of the mutations that made them a separate species.
5. Inbreeding must have occurred in the early stages of population growth.

How can these contentions be justified?

Proposition 1. The human species probably began with a single breeding pair. This would follow directly from the rules for the formation of new species by a change in the parameters of the computing formula. Alternatively, it is possible that the same mutation took place in a number of individuals in the same area and at roughly the same time (the founder principle). Several species have passed through bottlenecks and such a human bottleneck was located with the discovery in 1987 of mitochondrial Eve,[21] a putative maternal ancestor of every living human who lived in Africa between one hundred thousand and two hundred thousand years ago. Evidence for and against such bottlenecks continues to be discussed, but they are in principle quite possible. There may even have been a series of bottleneck events.

However, more direct evidence is available. The study of the shape of skulls of Neanderthals and humans suggests that the difference between them may be one of only a few mutations, possibly even just one. According to the views of Daniel Lieberman[22] of Harvard, the crucial adjustment was that of the angle at the cranial base of the skull, which in humans is more acute by about 30 degrees than in the Neanderthals. This results in a skull shape in which the face and eyes are tucked under the brain, allowing more room for expansion of the frontal and temporal lobes. This in turn may have supplied the cognitive capacity of modern man.

Proposition 2. Both individuals of that breeding pair were the mutant descendants of the same precursor species, and

Proposition 3. Both individuals would have died without issue or their offspring would have been infertile had they not mated with one another.

In addition to the mitochondrial Eve hypothesis, there is a lesser-known but perhaps equally important chromosomal Adam hypothesis.

It seems likely that between 150,000 and 200,000 B.P. a single male individual carried the Y chromosome that is now common to every human male.[23] There is even evidence that chromosomal Adam and mitochondrial Eve lived in the same place and at about the same time, since Adam was probably also African. They may never have met,[24] but if they had, it would have materially increased the probability of successful mating leading to a new species.

Proposition 4. Humans displaced the precursor species largely as a result of the mutations that made them a separate species. There seems to be historical evidence that humans and Neanderthals coexisted for some thousands of years,[25] while Neanderthals are now generally agreed to be extinct.[26] Earlier speculations that the two species may have mated and that some modern humans are their descendants now seem contradicted by DNA from Neanderthal remains.[27]

Proposition 5. Inbreeding must have occurred in the early stages of population growth. Again, this follows from the mathematical evolutionary hypothesis according to which the species begins with a single breeding pair.

CIVILIZATION PRECEDING OURS

It may be that the tribes had a civilization in the preceding interglacial period, based on flint technology and warmed and fed by fire. Remains of artifacts have been found from fifty thousand years ago in deep caves known to have been inhabited by Neanderthals. It has been assumed that these were made by modern humans, and that the Neanderthals came by them somehow and carried them there, along with fire to light their way. But an alternative is that Neanderthals themselves, or another branch of hominids, created and used them.

Humans have from the first been prone to warfare, both against their own kind and against others. War has been waged by whatever means seemed appropriate in order to subjugate or, if need be, exterminate the rival group. This must have been so in the relations of the new humans and their Neanderthal ancestors. Competition for resources of food, water, or living space would have led to one-sided campaigns against the Neanderthals. True, a Neanderthal in hand-to-hand combat might have killed a human. But humans had mastered arts of war unknown and impossible of achievement by their more sturdy but slower and slower-witted rivals. Humans could use tools of warfare: clubs held and wielded with skill, or missiles thrown with deadly accuracy. They would have outwitted the Neanderthals in ambush and skirmish, with traps and deadfalls; they would have poisoned their wells with plants and rotting carcasses; burned their shelters; stabbed and hacked at those who fell injured in battle. All the basic arts of war must have been practiced by the first humans in their battles against the near-humans, and they would have triumphed completely, driving the Neanderthals from the prime lands, isolating them in remote areas where it was unprofitable to follow them, and marginalizing them when they did not wipe them out altogether.[28] This edifying campaign is a prototype of later ones in man-to-man combat, and we would do well to heed Heraclitus' message, stated here in the particular rather than the general form, that "war is the father of all, some he makes slaves and others free."

THE PLIGHT OF THE NEANDERTHALS

Neanderthals seem to have been fully possessed of many of the thoughts and feelings of modern humans. The ability to perceive the wonders of nature, to admire the beauty of their fellows, to feel the awe of religious experience and the fear of the unknown, all these would have been available to the mind of the Neanderthal and the new human, and perhaps in equal degrees. If they lacked permanent expression among the Neanderthals, it was because of a lack of the means of expression. Without speech, writing, and the skills of drawing, Neanderthals could not leave for posterity a record of these thoughts and feelings and we can only deduce them from what we know about their physiology and their few surviving artifacts. Many skills and arts were totally unknown to them. The ability to count depends on the left hemisphere's differentiating forms and distances. Poetry without language is an impossibility, except as a crude heaving form in the mind. Music, while not unknown, without the invention of the bow could not extend beyond a primitive wailing melodic line or a thumping on wood or earth.

Consider the fate of the Neanderthal mind, imprisoned in a skull too heavy to emit the thoughts within: able to conceive beauty but not to express it; able to think but not to communicate; able finally, after the emergence of humans, to witness beings who could do these things while it is forever denied the same opportunity. Such prehumans were a living tragedy indeed.

But while the Neanderthals could not express their thoughts about humans, humans could express their thoughts about the Neanderthals. It is probable that races of protohumans such as the Neanderthals survived in small numbers in some places contemporary with modern humans, and that even after they disappeared their legend still survived, in fairy tale and myth, as stories of dwarfs and giants.

SEX AMONG EARLY HUMANS

If a species is to increase in numbers beyond a single breeding pair, inbreeding must take place. In the case of early humans this inbreeding (or, as we should say, incest) would have been indispensable. However, contemporary genetic diversity suggests that later interbreeding with the precusor species took place. Initially, sexual relations would have been between Adam and his daughters or between Eve and her sons, and between their sons and daughters. It is unlikely that this would have been problematical in a moral or esthetic sense. In later times, many societies accepted incest between privileged individuals. The Pharaohs were compelled to marry siblings, probably to prevent succession, and with it power, from passing out of the hands of the family. Early humans, having committed this divine (because necessary) act, shut the door on the deed and it became thereafter a societal taboo. The emotional effects of incest on the individual can be powerful, and this would have led to the creation of laws to protect against their effects. So a strong incest fear remains in our culture, although, paradoxically, there could never have been such a culture without it. Very likely, in the first days of humankind, incest was positively helpful, cementing the bonds of the little band even more strongly and adding intimacy to intimacy without prohibition in a way that can never quite be achieved in modern times.

SELF-SIMILARITY AND MORPHOLOGY

One of the strongest arguments in favor of a mathematical view of morphology is the similarity between their various parts that biological organisms display. This is an instance of the more general property of self-similarity that is characteristic of fractals. One part of the body can easily be mistaken for another, a fact often used by artists, designers, advertisers, and others to hint at what it may not be desirable or discreet to reveal. In close-up, the curve of a hand can resemble that of the torso; the shape made by the hand and the wrist can seem the same as that made by the leg and the trunk; the cleavage between the breasts is like the cleft between the buttocks. In a famous surrealist painting[29] by Magritte the resemblance between a frontal view of the naked body and a face has been deliberately exaggerated, highlighting the underlying similarity of form of one part of the body to another.

Typically, particular cases of self-similarity have been attributed to natural selection, as Desmond Morris did in his book *The Naked Ape*.[30] His explanation of the breast/buttock resemblance is that reminiscence of rear-entry mating is

9

ENTROPY, INFORMATION, AND RANDOMNESS

All things are an equal exchange for fire and fire for all things.

(FRAGMENT 90)

RANDOMNESS RECONSIDERED

In the first chapter I argued that the concept of randomness is central to the present worldview in science and that in order to progress we must change that view. This means we must reexamine the concept of randomness. In popular science the ideas of randomness and chance mean something that is the result of a process that has no pattern and is therefore completely unpredictable. But neither of these ideas—unpredictability or lack of pattern—is part of the mathematical definition of a random distribution (which is simply one that has an equal probability of occurrence of all classes of its elements), and it is debatable whether these popular notions can be deduced from this definition. To see why, we must look at both the differences between these generally widespread conceptions of randomness and what the definition implies.

The operations of chance are a constant source of wonder to gamblers, big and small. The concepts *fair* and *random* do not always coincide. If a coin is tossed into the air ten times, it might seem fair that on average it should come down five times heads and five times tails. In fact, on any given set of tosses this is less likely than some other distribution. The seemingly fair outcome, the one favoring heads and tails equally, will occur slightly less than one time in four. But the penny is still considered by most people to be fair in the long run.

One of the implications of the definition of randomness as it is traditionally conceived is that the events in a random sequence are independent of one another. Within such a sequence, randomness is both the reason why you get

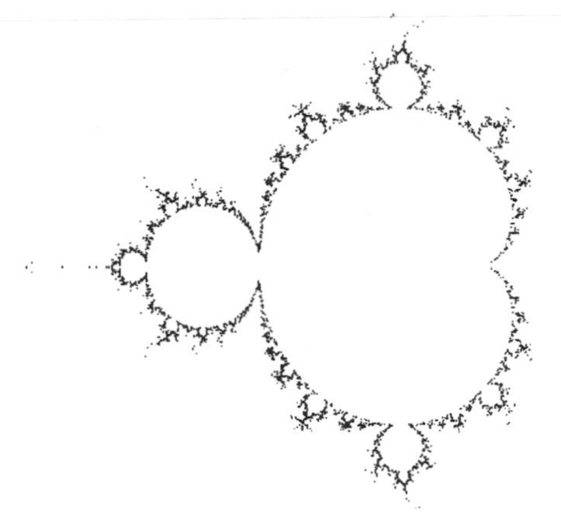

FIGURE 4.8 The Mandelbrot set.

esting blobs. Indeed, they were thought at first to be just the product of a computing error of some kind. When explored, however, this figure turned out to be a fractal of unlimited complexity. Iterated only 150 times, it looks like figure 4.8.

This figure came to be called the Mandelbrot set. One view of the Mandelbrot set is shown in figure 4.8. This is called an escape diagram and is the boundary between the points that stay inside the set after 150 iterations and those that fly off to infinity. The Mandelbrot set is certainly a fractal, and it shows a high degree of self-similarity when magnified.

The Mandelbrot set enjoyed a great vogue for several years. If the formula is iterated further, and if the resulting escape diagram is enlarged, the boundary takes on great complexity. If colored according to the iteration number when the points escape, this formula, one of the simplest possible in a plane, produces flamboyant and yet subtle shapes of extreme beauty.

Immediately, this shape looks of great interest to the biologist. Some people called it "the beetle" and it does indeed seem to have a head, thorax, and abdomen, as well as rather a lot of legs, or places where legs might grow. This resemblance between the Mandelbrot set and living things can be made even more explicit and we will look at how to do this in chapter 6. The Mandelbrot set is based on the simplest of iterative formulas and yet possesses potentially infinite complexity. Life is a process of growth, cell-division, and reproduction. Is it possible that in fractals the iterative nature of mathematics and the iterative nature of biology meet?

runs of luck and also why, if they are only luck, you cannot use them for long. Sooner or later in a random sequence of coin tosses there will be a run of, say, ten tails; but the probability of a coin coming down tails if it has fallen tails ten times already is still the same as it was at the start of the sequence of throws. So you cannot bet on the next toss coming down heads with any greater likelihood of winning. The belief that the probability of future events has altered as a result of past events is called the *gambler's fallacy*: there are two complementary forms of this. Suppose the penny has fallen tails ten times already: the first is to bet on heads on the grounds that after such a long run of tails, heads is long overdue. The second, slightly more rational, is to bet on tails on the grounds that the penny is not in fact fair. (It seems worth remarking that events in the world must be nonrandom; otherwise scientists are guilty of the second gambler's fallacy whenever they try to predict the future from the past.)

LIFE'S LOTTERY

When a British state lottery was introduced in 1957, people were invited to buy investment bonds that would be put into a draw, the winners getting sizable cash prizes. In order to make the enterprise profitable, the bonds, known as Premium Bonds, were used as stakes in a gamble but earned no interest, and indeed suffered the loss associated with inflation, then running at around 3 percent a year. In widespread media coverage people were told about the technology of the lottery. A machine called ERNIE (Electronic Random Number Indicator Equipment) was used: it worked by amplifying noise in an electronic circuit and harnessing it to drive a number generator (a suggestion first put forward by Alan Turing). The numbers produced were deemed to be random. The use of this term was meant to give some confidence to investors that the whole thing would be fair—that, for example, in the long run no more than the expected proportion of winners would be residents of the county of Essex.

The prizes to be won by Premium Bond holders were not very large compared to soccer pool winnings—for many years the major source of the big gambling prizes in the United Kingdom—and perhaps for this reason the media and the public gradually lost interest in Premium Bonds. When the National Lottery was introduced in 1995, the people responsible had learned from past mistakes and they were determined to make the lottery as attractive as possible. They did this mainly by making the prizes very large indeed. Jackpots of over 20 million pounds could be won and tremendous public interest was aroused in how to win them and who was winning. Remarkable coincidences soon appeared. For example, two of the early jackpot winners came from the same town—Blackburn—and lived within a few streets of one another. (The Blackburn soccer club was then at the top of the league, and many people concluded that Blackburn was a "hot spot," a lucky place to live.) Numerological studies were made of the frequencies of numbers that had been drawn. Some numbers turned up with unexpected frequency: the number 13 occurred only five times in the first 72 weeks, while the number 44 had occurred as many as fifteen times in the same period. According to the expectations of chance, both ought to have occurred ten times; this might lead a gambler to the fallacious conclusion that the number 13 was indeed an unlucky one.

To the public, these kinds of outcomes were not identified with something wholly random, although the promoters of the lottery maintained an apparently scrupulous fairness in number selection. Each week a different machine for mixing the numbered balls was chosen from a set of six machines and the set of balls to be used was chosen from several different sets (again by lot).

Far from being contrary to the definition of randomness, runs of luck are to be expected. In a long enough sequence of random numbers from zero to nine, it is expected that a run of ten 1s will occur and should (one cannot say must) occur sooner or later. If the sequence is extended long enough, even longer runs of 1s would be expected. In theory, though it will never happen in practice, a sequence of a hundred, a thousand, or even a million 1s should occur. Indeed, if this were not possible, the sequence would not be random. (Tables of random numbers can be bought and used by scientists to randomize some experimental design factors. What has just been said implies that, if you are shown a row of ten 1s, you cannot say whether it is part of a random sequence or not. So, in theory, a table of random numbers could be published, all of which were identical, although not many copies would be sold to experimental scientists.)

There is nothing necessarily nonrandom about a lottery that produces neighboring Blackburn winners on two successive occasions, or even several times running. Indeed, the same person might win two weeks running without the system's being unfair in the sense of nonrandom (though it would certainly *seem* unfair). A random system is capable of producing such a result. So randomness must not be confused with fairness; sometimes the operations of randomness may seem very unfair.

Randomness is a very useful concept when it can be applied in the real world. Its applications, apart from gambling, include agriculture, psychology, and the encryption of secret information: nevertheless, the nature of randomness has proved elusive.

A dictionary defines random as "having no particular pattern, purpose, organization or structure," and the word is derived from *randon*, which means a headlong dash or gallop, and is related to the French verb *randonner,* meaning to wander. This notion of uneven motion expresses the underlying expectations of randomness well: a progress involving unpredictable changes in direction.

THE RANDOM WALK

The true definition of randomness is mathematical, and like all such models we would expect to find it embodied in natural processes. This mathematical concept of randomness is derived from the idea of a *random walk,* the path taken by a drunken walker in a pathless field of limitless size. Various questions can be asked about such an ideal being. How far will he have wandered away from his starting place in a certain time? After a certain time, what is the probability of arriving back at his starting place? How often will his path intersect itself? To define a random walk, we assume that there is a sequence of steps and that it is equally probable that a given step will be in any particular direction.[1]

The concept of the random walk has been developed and applied to many kinds of situations. The question is, how do we know that the mathematical model applies to the real world? Since the definition of randomness depends on probability theory, we have to test the appropriateness of probability theory for the situations to which randomness has been applied: it must be a matter for experiment rather than mathematics to determine whether it fits or not. This involves counting frequencies of events to see whether they fit the predictions of an idealized random walk. (It also depends on various epistemological assumptions, particularly the assumption that future events will follow past trends.)

To the assumption that probability theory is a good description of the world, there are two kinds of objections, one purely mathematical and one empirical. The mathematical objection is that a truly random sequence must be infinite in length, otherwise the probabilities it generates will not be equal, so in a finite world there cannot be any such thing as a set of truly random numbers. The empirical objection, or objections, come from studies of randomness, which have all come to the conclusion that it is very hard, if not impossible, to find a genuinely random distribution in nature.

A textbook[2] on the applications of random noise written in the nineteen fifties begins: "There are many examples of random processes in nature, some of which are meteorological phenomena, thermal noise in electrical circuits, Brownian motion of particles, vibration phenomena and certain economic fluctuations." Since those words were written, there has been a huge development of nonlinear science and its ramifications, particularly in the field of chaos theory. Studies of the phenomena this book mentioned now suggest that at least two, and possibly three, of these examples of supposedly random processes can be modeled in a deterministic way. It seems that what at first appears to be random may not prove to be so on closer examination. We must seek deeper for examples of randomness.

Scientists bet on the outcome of their theories because their careers depend on them, but spies often depend on the secrecy of information for their very lives. For the secure encoding of information, a representation must be found that cannot be read by the other side. This depends on finding a sequence of symbols that is completely random to represent the message; otherwise the code-breaker can use the nonrandomness of the message, perhaps in conjunction with a guess as to its likely content, to force an entry. It has proved very difficult in practice to achieve this; there seem to be no unbreakable codes, because there are no completely random processes in nature.[3]

One possible definition (though by itself it isn't sufficient) is that any sequence of numbers is random if any subsequence of it contains, on average, the same distribution of numbers or sequences of numbers. However, as we have seen, in this sense there can be no such thing as a random number in the world, since all sequences are finite in length.

A definition of randomness that became popular in the seventies is that devised by G. J. Chaitin, based on information theory. It follows. All numbers that can be written down can be regarded as capable of being produced by computer programs; for example, there is a program that will print the number 0.5; there is another that will produce the number π; and there is one that will print the decimal 0.142857. Even if there is not a formula to generate the number, a program can still be written simply to print the sequence of digits that comprise it—its decimal expansion. Now the amount of information in a sequence, whether it is a number or a computer program, can be measured by the number of symbols it takes to write it down, so the informational content of the program and the number it generates can be compared. For any program and its output there will be a relationship between the length of the program and the length of the number it produces. The degree of randomness[4] in a number is defined to be a function of the relation between the length of the number and the length of the program that produced it. A random number is one whose corresponding program is greater than or equal in length to the number it generates: if the program is very much shorter than the number it produces, then the output is nonrandom.

This definition of randomness has problems. The ratio of the circumference of a circle to its diameter is called π, a number that starts 3.141592. . . . A formula to produce the value of π is very simple to write, and using recursive notation, it can be expressed in very few symbols.[5] A computer program to produce π can therefore be written, using recursive programming methods based on such a formula, that will eventually produce more digits than are required to represent the program. No one, however, would assert that π was a number that was predictable or had a recognizable pattern.

Many different programs—indeed, an infinity of programs—will generate the same number. Which is the one that should be used for the comparison of lengths? The shortest perhaps. But how do you know which is the shortest? Many programs have been discovered for the calculation of π; perhaps a shorter one will be found someday. Would this alter the degree of randomness of π?

Defining randomness by the arbitrary nature of how a computer operates seems to evade the problem. There is no simple relation between a set of statements in a computer language and the operations carried out inside the machine. A computer program may be simple, but the operations carried out by the computer may be very long-winded. For example, "Print the square root of 2" is a short program, but the extraction of a square root takes many microinstructions and the duration of computation itself is of indefinite length. Is the sequence of digits in the square root of 2 random or not? There seem to be three possibilities here: (1) since the program required to produce the square root of 2 can be written using very few symbols, then by the preceding definition it is highly nonrandom; (2) but because the program is really much longer than it appears to be, the square root of 2 is less random than this would suggest; and (3) because the sequence of digits in the decimal expansion of the square root of 2 is completely unpredictable, it also appears to be completely random. Which is the correct answer?

INFORMATION AND RANDOMNESS

Problems like these make the Chaitin definition of randomness untenable in its original form. The information theory approach is nevertheless interesting and it is worth pursuing it further.

Modern information theory began with the work of Shannon and Weaver.[6] Their approach was based on the idea that something contains information if it surprises us: information is essentially the same as news.[7] This capacity to surprise depends on the likelihood (or unlikelihood) to us of the symbols of the message occurring. Therefore, the amount of information in an array of symbols can be given a quantity that is a function of the probability, as understood by the receiver, of each symbol occurring.

For example, if we are using the normal English alphabet, the letters have different probabilities of occurrence in normal speech or writing. The letter *e* is the most frequent, then *t*, and then *a*. The further down the frequency scale we go, the harder it is to be certain, but from this knowledge of frequencies we can assign different information values to different letters.

There are, moreover, sequential dependencies that lead to one letter or group of letters being followed by another with greater than expected probability. For example, if we are sent a three-letter group of symbols in English beginning *th*, we can be sure, excluding dialect or archaic speech, that the third letter will be *e*. This means that a meaningful letter sequence like *th* together contains less information than *t* and *h* do separately. If the message contains a lot of information, it will be more unexpected, its symbols will have lower probabilities of occurrence, and the sequence will be more random. So information and randomness seem to be linked.

The idea of randomness also seems to fit in very well with the modern concept of chaos. Chaotic motion is in many senses random; it is unpredictable, it is apparently disorganized, and it is without an evident pattern. Is there any difference between randomness and chaos? In fact, the appearance of randomness in chaos is an illusion. Chaotic systems are well-ordered and their results are predictable to some extent, while the essence of randomness is unpredictability, except in the statistical sense. In practice it is possible to distinguish between random and chaotic systems by means of correlation analysis.[8]

The trouble with this approach is that the amount of data we need increases proportionally with 10 to the *n*th power, where *n* is the correlation dimension. For example, if we want to detect a correlation dimension of 3, we need 10^3 or 1,000 points; if the correlation dimension is 5, we need 100,000 points, and so on, so for large dimensionalities we rapidly run out of data. As a test of randomness this is not a very good one; however, it does indicate that randomness shades off into chaos of a particularly complicated type.

IS 1 A RANDOM NUMBER?

Consider an infinite sequence of 11111. . .. This, as it stands, conveys very little information, and if it is the only thing we know about the situation, we might assume that 1 was the only possible digit that could occur, and that therefore the sequence told us nothing. If only the digit 1 could occur, then this sequence would be empty of information. If we see 101010. . ., we may be justified in assuming a binary system in which we can have the digits 1 and 0, and in concluding that the array has two bits of information, repeated over and over again.

The number $\pi = 3.1415926535\ldots$ contains a lot of information, potentially an infinite amount. So it is possible to say that a random number like π—random in the sense that it consists of an unpredictable series of digits—contains an infinite amount of information. We usually think of a random number as containing no information because it has no particular pattern. But information theory seems to indicate that a random number contains a very great amount of information. Can we square these two understandings of randomness?

It is possible to see the problem in a new way by thinking about arrangements of objects that are apparently random but are really orderly. Take a chessboard: when the pieces are lined up at the start of the game, they look orderly and nonrandom. In fact, they convey little information, because the game has not yet begun and they resemble the start of every other chess game. If we look at the pieces after thirty moves, they will appear much more jumbled and, if we did not know the rules of chess, it might look as though they had been placed at random on the squares of the board. The arrangement is actually highly constrained. The players do not move at random but with their position and their future advantage firmly in mind.

Or consider cars passing you in the street. If you list their registration numbers, you will get an apparently random sequence. But when the cars left the factory, they probably had numbers assigned in ascending order. They were then driven around and have now arrived in the street in the order you see. If you knew their movements since acquiring their registration numbers, you could account for the sequence you observe. It seems to be random, but it potentially contains a large amount of information about where the cars have been. In fact, the sequences of numbers of cars, checkroom tickets, and so on is often not at all without order. Carl Jung used to record such coincidences, and so did the biologist Paul Kammerer.[9]

To return to the example of the Rubik's cube in chapter 1: after several twists following a certain rule, it assumes a random appearance, but it is fairly easy to restore it from such a random setting if you know the rule that produced it. However, if you were to give the cube to someone else and ask them to restore it, they, not knowing your rule, would have to use an algorithm for restoring the cube: the most efficient algorithm known would need more than eighty separate twists to get the cube back to its original orientation. Now, in fact, a Rubik's cube can never be more than twenty moves away from its restored position, but no one could possibly restore a cube in that number of moves. Why not? Because they assume that the cube is randomly arranged, when it is in fact arranged according to an unknown sequence of moves. Rather than seeking perfect knowledge, people adopt the brute force approach and apply an algorithm. But the key is there, just out of reach.

We can see from these examples that sequences that look random are often far from being disorderly; such sequences contain a certain amount of information. Are there any truly random sequences? It is difficult to say whether there are unless we have an adequate definition of randomness, and there isn't one at present. We have seen what is wrong with the existing approaches: they don't meet the needs of the concept of randomness. Is there another approach that is more adequate?

Random numbers are used by scientists to simulate conditions in the real world, in particular, when they want to plan an actual experiment or simulate one that might have taken place. They are used by statisticians to arrive at the significance level of experimental results on the same basis (what would be expected by chance). But what happens in random number tables and what happens in the world may be very different things.

Many people, scientists and nonscientists, believe that life has come into existence and taken up its present forms through the operation of random processes. But we can see that randomness is in itself a very complicated matter. It is not possible to give a definition of randomness that is other than a mathematical one: no such thing as a random sequence of events can exist in the observable world, which is finite. There is good reason to believe that underlying apparently random events there may be order, if only we could discover what that order is.

ENTROPY AND ORDER

One of the biggest scientific puzzles posed by the existence of life in the universe is how it can produce and maintain order in a world that seems inevitably to tend toward the breakdown of order. It is a commonplace of experience that disorder increases unless you do something to prevent it. The house gets dusty; the dishes have to be washed every day. Life often seem to be one long and often vexing round of arranging objects and tidying up messes, either your own or other people's, and it seems as though disorder is constantly being created by some apparently ineluctable force.

Our intuition that disorder tends to increase is borne out by physics: the second law of thermodynamics can be variously stated, but a common version is that in any system, entropy (disorder) increases. The concept used in the second law, that of entropy, is a mathematical measure of the distribution of probabilities. It is defined so that a higher entropy value (varying between 0 and 1) is equivalent to a greater degree of disorder. This implies that in an untidy arrangement of objects, the entropy is greater than in a tidy one. For example, if you keep all the spoons in one drawer and the forks in another, the entropy is less than if they were jumbled together in the same drawer or if they were not kept in a drawer at all. Entropy is a measure of disorder: its opposite, negative entropy, would be a measure of order.

The concept of entropy appeared first in the study of thermodynamics and came about because engineers were looking for a way to describe the transfer of heat in machines such as steam engines. This was important in the study of these engines and how to make them more efficient. There are many ways in which heat can be transferred between two places within a machine and still lead to the same final state, so what was required was a way to describe these heat-transfer processes—a way that would enable engineers to understand better what was going on. The approach that was taken was to divide heat energy into small packets, or quanta, and to consider the distribution of these heat quanta within the machine. Since probabilities are involved in such a situation, statistical techniques have to be applied to obtain valid predictions of what happens when an engine is using energy in the form of heat. One of the consequences of this kind of analysis is that the concept of entropy—probabilistic distribution of heat quanta—emerged. Three laws of thermodynamics were formulated; the second states that entropy inevitably increases. The first law states that energy cannot be destroyed, and the third that heat is never transferred from a cooler to a hotter body.

The second law of thermodynamics has become something of an icon in the scientist's gallery and has firmly established the idea that any orderly situation gradually yet inevitably becomes less orderly as time goes on. Not only is there a tendency for disorder to increase, but that tendency seems to be irreversible. About fifty years ago, when the second law gripped the popular imagination for a time, the idea of "the heat death of the universe" became fashionable. What this lurid phrase expressed was the idea that the whole universe was running down as heat energy became more evenly distributed, and would one day become a tepid soup of particles moving at or around some average velocity. When, countless millions of years in the future, this finally occurred, no more systematic change would be possible and the universe would have finally run down, like a clock whose spring has uncoiled. Without the motor of heat differences to drive events on, there would be, in effect, no more events. Since everything would be at the same temperature, it would not be possible to perceive objects, even were there any organisms to perceive them. As T. S. Eliot put it, "we all go into the dark . . . into the silent funeral, nobody's funeral for there is no one to bury"[10] or, in another poem, "This is the way the world ends, not with a bang but a whimper."[11] A bang implies the concentration of energy; but a too-even distribution of energy is just as surely a destruction, not merely of a few things, but of everything.

TIME'S ARROW

The second law of thermodynamics talks about heat and its distribution, but the principle it expresses—that disorder inevitably increases—can be generalized to describe any distribution of matter or events. An often-used example is mixing two liquids together: if you pour milk into a cup of tea and stir it, the milk will eventually disperse evenly throughout the tea, changing the color from dark to light brown. The milk and the tea are in an orderly state prior to being mixed; afterward they are disorderly. It is then impossible to recover the original separation of tea and milk; you cannot stir the tea backward and get all the milk back into one part of the cup. The second law of thermodynamics is thus connected to the idea of time: "time's arrow," to use the notion of the Greek philosopher Zeno, which moves in one direction only—in the direction of increasing entropy.

The great paradox of life is that it seems to defy this rule by maintaining orderliness in a universe in which orderliness tends to diminish. While things are in general distributing themselves in a more disordered way, life is continually building up order. Chains of nucleic acid replicate, produce amino acids, and from these, proteins and cells and somas are formed. It is true that individuals eventually die and that this brings to an end the orderly arrangement of the components of their bodies, but the race, and life as a whole, continue. Despite the advance of time, life maintains the orderliness of its constituents and, through evolution, it even appears to increase it. Since life seems to circumvent the second law, a lot of thought has gone into the question of whether there is a flaw in it. Perhaps the second law is not really a law at all but just a matter of probability and some clever enough mechanism, such as life evidently uses, is capable of defying it. Can the second law be circumvented?

It seems intuitively as though it ought to be possible to overcome the second law, at least in a limited volume of space or for a limited period of time. As a conscious decision-making organism, you can separate things that have become jumbled together. You can take the forks out of the knife drawer and put them in a separate one, but in order to do so you must use energy, and this will entail an increase in entropy somewhere else—in whatever mechanism provides that energy.

UNSTIRRING THE TEA

In the example of the cup of tea, there is no easy way to recover the original constituents, that is, the flavored water and the milk, and this is a practical difficulty. But supposing a way could be devised, say by evaporation and reconstitution, then such a method would require the addition of considerable energy to the system. The energy would very likely come from electricity and this has to be generated, perhaps by burning coal or gas. But in burning a fuel such as coal you in turn increase its entropy. So a decrease in entropy (order) in one part of the universe inevitably produces an increase in another. Considering the coal and the tea together as parts of a closed system, entropy is still not decreased, and the system as a whole even gains some entropy because of the leakage of heat to the environment.

It might be suggested that a way around this would be to get your energy for the evaporation process from sunlight, but this would depend on the fact that the entropy of the sun is gradually increasing. Indeed, the sun can be regarded as a vast well of negative entropy from which life draws its supplies of orderliness. The sun is radiating in all directions at once; the part of its radiation falling on the earth is our supply of negative entropy donated free by the sun. This is no free lunch however; if we want to use the sun as a source of energy, we must reckon into our calculations the entropy gained by the sun in the process.

Another alternative for unscrambling the cup of tea would be to "stir it backward." We all know that this is impossible with milk and tea, largely because two sets of molecules are bound together. But if the right kind of backward stirring could be contrived for a mixture, perhaps it might be possible. In fact, it is possible. If you put a dispersible aspirin tablet in a little water in a glass and stir it with a spoon, you will get a fairly even distribution of particles. (Because the aspirin is not soluble, the particles will remain whole.) In order to get them back together again, all that is required is to gently swirl the water in the glass, as people sometimes meditatively swirl their drinks. The aspirin grains come together in the center of the glass and could easily be removed. Rather than using a brute force method of restoring order, like evaporation, we are thus able to restore order by means of the same operations that produced the disorder.

This is cheating a little, because the particles have a different density from the water and so collect under rotary motion in the middle of the cup. But essentially what is required, backward stirring, can indeed be done if the circumstances are controlled well enough. If two liquids of different colors are put into the space between two concentric cylinders, and they are rotated relative to one another, you can see the two mixing until after a few rotations they appear quite well blended. However, simply reversing the rotation will restore more or less the original distribution of two separate colors. So it seems that you can reverse time's arrow, if it is done in a confined enough space.

INTRODUCING MAXWELL'S DEMON

This unstirring idea depends on the laws of physics being reversible. How nice it would be if we could reverse any event by reversing the operations that produced it. For some events this is possible, like turning a clock backward, because the hands are designed to go in either direction. Likewise, if we could find a way of stirring the tea backward, then we could neatly scoop the milk out with a spoon. What kind of mechanism would enable us to do this? One answer that has been proposed is Maxwell's Demon. The Demon is a small being in the cup of tea and is pictured as an energetic little creature with a minute tennis racket. He could be instructed to hit a particle of milk up to the top of the cup when it comes his way; when a molecule of water comes along, he should hit it downward. The end result would be that the two liquids become separated and all the milk ends up where it was poured in. If we could harness such a demon for our purposes, it might enable us to achieve order.

Maxwell's Demon is a figment of the imagination, yet he stands for a mechanism that might really exist. You may not be able to conjure up a demon in your tea, but you could have something along the following lines. Suppose the cup of tea is divided by a barrier into two parts, as in figure 9.1. Set into the barrier is a sliding door that can be opened and closed. The door slides back and forth as particles of different types are recognized (in practice, say, by an electric eye mounted on the barrier). Every time a milk particle approaches, the door is slid open, and it slams shut immediately to prevent any water from passing through. After enough operations of this kind (or you could imagine a lot of demons at work simultaneously) the two liquids would be separated into the original components. If such a machine could overcome the second law, the implications are profound; for by such means it would be possible to separate the faster-moving and slower-moving molecules in a liquid without doing any work. This would constitute a "free lunch" engine for producing heat apparently for nothing. Is such a device feasible in principle, and if it were, would it not overcome the second law?

You might think the snag is the energy cost of operating the door, because in our world, doors require energy to open and close. But by using a frictionless sliding door, this energy requirement can be reduced in principle to as

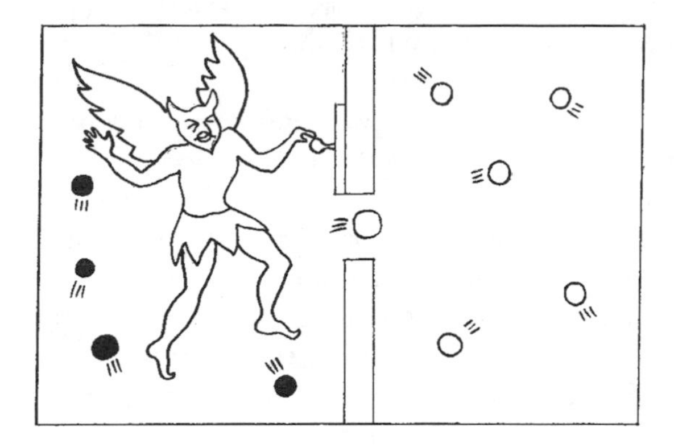

FIGURE 9.1 A cup of tea divided by a barrier into two parts, with demon. Drawing by V. G. Bird.

near zero as we wish, so this is not the barrier. The real difficulty comes from the decision-making mechanism. There must be a perceptual system, a magic eye of some kind, taking the place of the demon, to operate the door. This magic eye, which views the approaching molecule, must determine and make a record of what kind of molecule it is, fast or slow. Once the molecule has been dealt with, by sending it through or stopping it, its record has sooner or later to be cleared from the system's memory, so that the next molecule can be processed, and this erasure of the memory constitutes an increase in entropy in the memory device. The entropy lost by the molecule is gained by the memory of the door operating mechanism, which becomes the sink into which the entropy lost by the system flows.[12] This amounts to saying that you can arrange some part of the world in a more orderly way only if you disarrange another part and make it correspondingly more disorderly.

The only known influence running counter to this tendency of things to disarrange themselves is life. Living things stand in apparent defiance of the second law of thermodynamics, because while inanimate matter seems to become more and more disorderly in its distribution, living things alone retain their structure and even tend, by reproduction, to spread this order throughout the rest of the world.

MAXWELL'S DEMON LIVES!

How, it must then be asked, is it that life can somehow overcome this apparently universal tendency for disorder to increase? Life seems to be able to perform a balancing act of maintaining order in a world that generally tends toward the breakdown of order. What is the explanation for the ability of life to at least cheat, if not change, a law of physics?

Astonishingly, a demon like Maxwell's does exist and is one of the most useful beings in nature. An experiment analyzed by Eggers[13] is illustrated in figure 9.2. A container is divided in two by a board with a window near the bottom. Each half initially contains an equal number of steel beads and is vibrated at 10 Hz at 1.3 *g*. After a while the beads redistribute themselves so that slow-moving beads are mainly in one half, resting on or near the bottom, while the fast-moving beads are mainly in the other chamber. This is not Maxwell's Demon although it looks rather like it, because the window is open all the time, rather than being opened and closed, and the second law is not violated. The way it works is this. To begin with, the situation is symmetrical, but asymmetry develops; more beads collide in one half than the other, producing a difference in environments. In what is to become the slow-moving half of the container, faster beads collide more frequently and collapse to the ground, losing energy to the bottom of the container. There they will form a cloud of slow "molecules" that will drag others down to the bottom too. Beads are constantly jumping through the window, so if they find themselves in that half, they will be slowed down and join the slow-moving beads. Beads jumping the other way will join the fast-moving population in the other chamber. So a separation of order from disorder has taken place by purely passive means. No demon need apply!

This experiment is an example of a noise-induced transition—an apparently universal phenomenon. It arises out of the property of noise that it occasionally contains exceptionally large signals, and these can be used to cross a threshold. Imagine a pendulum working against a ratchet to drive a clock. Noise is applied to set it swinging, but not enough to work the ratchet. If the noise is sufficiently strong, then a large enough signal will occasionally push the pendulum past the ratchet; the clock will advance erratically, but it will surely advance.

Another closely related phenomenon due to noise is stochastic resonance,[14] a process discovered in recent years that is a way of getting order from apparent disorder. *Stochastic* means a process involving a random variable, so we are back at square one as far as definitions go. However, the recent recognition of the phenomenon represents something new. Stochastic resonance is a process by which a weak periodic signal can be amplified in the presence of noise and it occurs in many systems in domains ranging from the neural to the climatic.

Stochastic resonance is a process that is helpful in a number of ways: first in perception, where added noise can aid signal detection. A noisy background is not usually conducive to communication. If you want to say something to

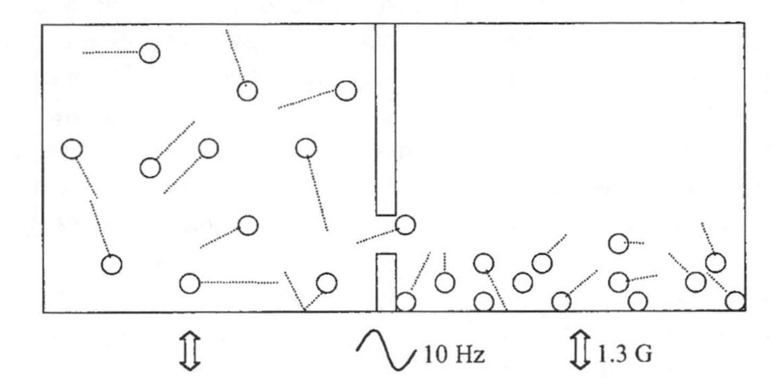

FIGURE 9.2 A container divided into two halves.
[After J. Eggers, "Sand as Maxwell's Demon," *Physical Review Letters* 83 (1999): 5322–5325.]

somebody, it isn't helpful to have to shout at them over the noise of a brass band. Noise when added to a signal usually makes that signal less detectable and the signal-to-noise ratio is the classic measure of detection. Adding noise to a time series washes out all the fine structure usually associated with the operation of chaos. But it seems that noise of certain kinds can actually make the perception of a signal easier, not harder.[15]

Stochastic resonance also makes energy available to the basic operations of the fibers that enable living things to move—the actin filaments.[16] The way in which this mechanism operates is not yet fully understood but it seems that occasional spikes of energy stand out against the background, allowing the ratchet to work.[17] It is as though we move against an environmental background noise that is not simply noise.

However, we are assured, nothing is lost before the second law—and this is as it should be; for in sequential operations information is not lost, just transferred. Stochastic resonance is a free lunch only for the diners at life's counter. For the world as a whole there is no new information.

LIFE AS INCREASING DISORDER

The preceding examples show randomness as rather different from what is imagined, and its principle as applied to life has frequently been misread. Life is assumed to be orderly, mainly because it has orderly outcomes. Birds build nests; people arrange cutlery drawers. It is easy to make an intuitive leap from such observations to the conclusion that life is essentially an orderly process, when all that has really been examined is some of the habits of living things, not the nature of life. Because the effects of life may be orderly it is a mistake to suppose that the process of life is itself orderly. Life, as I shall try to explain, is moderately disorderly and grows more so with the passage of time.

Let us look again at what is happening in the examples we discussed. In the first, the two liquids, milk and water, are not particularly organized to begin with. Because one milk molecule looks to our eyes much like another, it may seem that the milk is a consistent, evenly distributed, and therefore orderly substance. But all liquid molecules are in a constant state of motion and are always rearranging themselves, so their apparent uniformity should not be mistaken for regularity. What is meant by there being more order in the initial state than in the final state is that to begin with, there is milk here and water there, while afterward, there is milk and water everywhere.

Consider again the cooling down of a liquid to the temperature of its surroundings. To begin with, there is a population of warm (faster-moving) molecules in the liquid and another population of mixed warm and cool (slower-moving) molecules in the environment. Energy is lost in the form of radiation or convection: in the former case the molecules become less excited, in the latter case the faster moving ones escape from the liquid. In either case there is a greater distribution of molecular speeds in the liquid when it cools down. The heat energy in the liquid has become more disordered in its distribution.

Using the concept of randomness, as discussed above, we could say that the milk becomes randomly distributed throughout the tea, and that the energy in the molecules becomes randomly distributed as the liquid cools. Now, as we have seen, randomness is another term for informational complexity; so what happens in both these cases is that informational complexity has increased.

Informational complexity is similar to what is defined mathematically as entropy, and the second law of thermodynamics states that entropy can never decrease, it always tends to increase. Heat is never transferred spontaneously from a colder to a hotter body; it is always the other way around. Life is no exception to this process. The paradox then is not how life manages to evade the second law of thermodynamics, but just how entropy is used by living things. If life is complex information, why does the complexity not lead to disorder, as it does in the cases just discussed?

I think the answer is that, in scrutinizing the arrangement of matter or heat energy in systems and analyzing it according to the measure of entropy, we see the details but we miss the picture. The second law of thermodynamics is purely statistical, and is not like other scientific physical laws such as the laws of motion in classical dynamics. It tells us about the probability of certain arrangements arising in the world, but it does not tell us anything at all about *how* they arise. Its basic assumption is, indeed, that they will arise randomly, so it is not surprising that the popularly understood meaning of the second law is that things become more random over time.

Real-world systems evolve in a way that brings about their overall organization. This applies not just to life, but also to systems that are, strictly speaking, nonliving but have some of the characteristics of life. I have used the term *zoic* to denote systems moving toward the life-like in their degree of organization. Zoic systems have the characteristic that they tend to maintain overall order. But if we look at how the system as a whole is organized, we can see that the disorder that apparently increases on the small scale is lost in the order imposed by the system parameters. Disorder only increases if we ignore these global orderly tendencies. The increase of entropy is what happens on the whole, that is, most of the time, though occasionally and locally it can decrease. Entropy increases because individuals (particles, cells, people, or indeed any unit you may choose) that are in an orderly arrangement will tend after a time to move into a less orderly one. Suppose that there are a number of soldiers on parade. The entropy of this arrangement is relatively low, because it is unlikely to arise by chance; parades have to be arranged. When the parade is dismissed, the soldiers fall out and go their separate ways, so the entropy, the probability of finding the new arrangement of the men by chance, increases. Here I am assuming that one person goes to the canteen, one to the stores, one to get his hair cut, and so on. If this happens, as it usually will, then the neat parade becomes a confused scene of criss-crossing paths. This is equivalent to the increase in entropy that we have seen happening most of the time.

But it may be that this is not what happens. On some occasion the men may all march off in the same direction at once, in which case the entropy will remain unchanged. Why such a thing might occur doesn't matter; what is important is that although it can happen, it almost never does, so entropy almost always increases. On the whole, the arrangement of any set of things, whether they be soldiers, particles of soot, or motor cars, will become increasingly complex and apparently disorganized with time, so entropy will, on average, increase. And this increase in entropy is not the result of a physical law, but is purely a statistical matter; it is what is most likely to occur.

I have said that for short periods of time and in small localities the reverse will occur and entropy will decrease. This also follows from the laws of statistical mechanics. Every now and then a gambler will have a run of good luck and make a lot of money in one evening (though he will probably dissipate his winnings the following night). Similarly, it is possible that the cup of tea might on some occasion spontaneously become unstirred, so that the milk rises to the top again. But this is very unlikely ever to happen. What is far more likely is that there will be localities, small ones admittedly, in which a greater number of orderly elements than before will appear over a short interval of time. This would not violate the second law, because the overall statistical tendency is for entropy to increase. But nowhere is there to be found a law that says that entropy must always and everywhere increase—indeed such a thing would itself violate the second law! If entropy always increased, then one could, for a given locality, predict that a less orderly situation would certainly come about. And this would in turn mean that locally there was a higher degree of organization than there ought to be.

This paradox, that entropy must change in an uncontrolled way, for if it is a controlled way, entropy is not really increasing, is at the heart of the matter. For what we are talking about here is the operation of the laws of statistics, not the operation of a physical law. But need statistical laws really govern the universe, or our locality within the universe? Is there not something rather unsatisfactory about statistical rules, which are only generalizations, when true knowledge would serve us so much better? Is it possible that the increase of entropy, while it appears to be the outcome of a random unpredictability, is really hiding the operation of highly complex, yet perfectly orderly and deterministic, processes?

ENTROPY AND DIASCOPY

Let us return to the example of the game of chess. At the start of a game the pieces stand in rows; they look neat and orderly, as indeed they are. The entropy of this arrangement may be assigned the value zero, because it is certain that this is how the pieces must be before the game starts. Now suppose that forty or fifty moves are made by both players. By this stage of the game it is quite likely that none of the pieces is still on its original square. The situation no longer looks tidy, and to a non-chess player it might even appear to be entirely random. We can say that entropy has been added to the game.

But, if we know the moves, we can trace the path of each piece back, if need be, to its square at the start of the game. Some systems of chess notation are not adequate to this task, because all they do is describe the square the piece ends up on (plus what kind of piece, if more than one could have moved there). In the present task this leads to problems, because you cannot tell where the piece has come from, nor, if there has been a capture, which piece was standing on the square before the move was made. Some better system will have to be used, like the so-called Continental notation, which gives the start and end squares for the move, with additional information stating the type of piece captured, if any, so that it can be made to reappear when the moves are run backward. But, given such a way of recording the moves, it would be possible to take any board position, together with the history of moves so far, and run the game backward until you arrive at the starting position. Now since the starting position represents an entropy of zero, and since it is certain that this position can be reached from the present one, the entropy of the present position must also be zero.

This point can be seen in a different way by applying what may be called *diascopy*, or seeing through. A window can be seen through; so can a telescope. Although the window and the telescope are media through which information must pass, they are transparent media; that is to say, the information in the scene—the relationship between the parts of the picture—is unchanged after it has passed through them. In an analogous way the moves of chess can be made transparent so that the initial position can be "seen" from the final one. The concept of diascopy is rather like that of indirect addressing in computing. Instead of referring to a piece of data by its address, we use another piece of data that directs us to it.

To return to the game of chess, suppose for a moment that the pieces are living beings—the knight is a little knight, the pawns are peons, and the king is a king—and then imagine that a set of optical fibers that they carry connects the pieces to the squares they stood on at the beginning of the game, so that they can see them at any time. Now, when they move to a new square, they can still see their point of origin. No matter how they tangle the fibers, they will always see the squares on which they originally stood, and the fibers can be untangled simply by reversing the order of the moves. The same thing can be done, logically rather than physically, by applying the moves of the game in reverse. Inverting the moves is like untangling the optical fibers that connect the pieces to their original starting places. Viewed through this system of "logical fiber optics," the pieces can be imagined to be still on their original squares. Seen in this way, the arrangement of pieces still has zero entropy.[18]

This surprising conclusion makes it appear as though entropy can never increase. If such an argument were taken to its logical conclusions, it would imply that the entropy of the universe has not increased since the Big Bang! But this is so only if we neglect the history of the situation we are examining. In fact, there has been an increase in entropy, but as we know, the entropy can be reallocated from one part of a system to another. Taking the moves and the board positions together as one system, the second law applies. If the entropy of the chess board (or the universe) is artificially maintained at zero, the entropy that would have been gained by the board is absorbed by the moves, since the moves themselves represent unlikely sequences of symbols. Libraries could be filled by the transitions involved in getting the universe from one state to another state a fraction of a second later. These strings of symbols would represent positive entropy, since they are themselves improbable.

Another reason why entropy must be considered is that we rarely if ever have enough information to apply diascopy. If we have complete knowledge of a situation, it is true that the entropy viewed diascopically is always zero. In practice, we rarely have such information. But if we understood the rules governing the soldier's actions just as we understand those of chess, then the entropy of the soldiers alone could always be viewed as if it were zero! They might as well be standing on the parade ground once again. Doctor Johnson, when a friend boasted that he had traveled by cabs numbered 100 and 1 on the same day, said "Sir, the chances of those numbers are exactly the same as those of any other two numbers." Suppose the friend had traveled on cabs numbered 1 to 100 on one hundred successive days. The same would still be true. He has to travel by some cab, so each arrangement of cab numbers is equally as likely as any other. Viewed in this way, any arrangement of matter is as probable as any other.

THE MANAGEMENT OF INFORMATION

What living things do is preserve the degree of entropy or information that they have, not allowing it to increase beyond a certain point. Life is the management of disorder, not its elimination. As evolution progresses, living things become more complex, contain more information, and hence, in a sense, become more disorderly. But if this disorder is pushed too high, as it is in diseases such as cancer, then the organism ceases to function. It is this maintenance of pattern that is the key to the survival and relative permanence of living things. As Victor Serebriakoff points out in *The Future of Intelligence*,[19] if you take two horses, one alive and one newly dead, the dead one will have changed a lot more in a year's time than the live one. This makes the point that the essence of life is stability, or as I have tried to show in chapter 5, *chaostability*; and it is the forces of nature that are constantly trying to change that stability into disorder. However, we must be careful in assuming that because life does not change very much, it can somehow circumvent the laws of thermodynamics. The second law implies that order cannot flow from a less organized into a more organized system, but it does not follow that the reverse must happen. It is possible to hold the degree of order in a system constant at least for a while; this is what constitutes the miracle of life.

I have said that as evolution continues, living things become more complex. This is almost a truism, but it can be reinterpreted in a number of ways. We could say that organisms were becoming more complex as they evolved; we could say that they were achieving a state of greater entropy; or we could say that their information content was increasing. We could even, loosely, say that they were getting more random in the organization of their cells as they evolved. All these statements would be equivalent. What they all imply is that the net entropy of species increases with time.

There are some examples of evolutionary change where this does not happen; sometimes species evolve from the more complex to the less complex.[20] This is like the occasional eddy that runs counter to a current. In the cup of tea, there will occasionally be some pockets of milk, but they will be small and temporary only. Nevertheless, organisms as groups of species evolve toward greater complexity. What about the individual organism? As it grows and matures, it too becomes more complex, so its entropy increases during development, as shown by the iterative model discussed in chapter 8. Entropy inevitably increases for the individual as well as for the species, more quickly at first while development is taking place and then stopping. So, far from overcoming the second law, life over both long and short time scales, at the individual and species levels, exemplifies it.

FOOD AS INFORMATION

The first requirement for life is food, so it would not be surprising if the subjects of food and entropy were connected. The physicist Erwin Schroedinger in *What Is Life*[21] explored some of the implications of the concept of entropy for biology. He suggested that there is a basic requirement of living organisms for low entropy. This negative entropy, as he called it, was necessary for the maintenance of the creature's organization.

This artificially lowered level of entropy was achieved through the apparently unlikely mechanism of diet. He pointed out that food is not simply disorganized matter, but is itself invariably the product of other organisms. It is not possible to get any nutritional value from inorganic chemicals, which are essentially disorganized. We can only survive by eating the tissues of plants and other animals, which, being highly organized, possess low entropy. Of course, food gives us not only negative entropy but also energy to be used by the muscles and by the mechanisms that maintain body temperature. However, Schroedinger said, the main reason for eating is not the energy it provides, but the negative entropy we acquire from it. As he pictured the process, food is taken in and the parts that have lower entropy are used. The higher entropy components of food are excreted. Those are the parts that are more complex and disorderly than those from which we can extract negative entropy for our own use.

This interpretation, which at the time was strikingly novel, has become almost a standard item of belief among physicists, though it was opposed by many biologists (in the 1940s) and is still rejected by some as a sort of physicists' heresy. But to look at eating in this way suggests that what organisms are doing is trying to prevent too sudden an increase in their entropy. In this view, life is always moving toward a decrease in entropy. In order to do this, it takes in stores of negative entropy by way of food, the eventual source of which is, as Schroedinger points out, the sun. Stars are immense stores of negative entropy, which date from early in the history of the universe, and ultimately from the Big Bang, the point of minimum entropy.

I want to suggest a different view. *Entropy* is the Greek translation of the word *evolution* and this etymology provides a clue to the link between entropy and evolving life. We have seen that entropy is the distribution of information, and we also know that life is based on sequences (e.g., nucleic acids and protein arrangements) that contain information. The nucleic acids—DNA and RNA and their variants—carry information that is required by the organism to lay down its pattern of development. Protein sequences prescribe the nature of the substances that make up the soma of an organism. These nucleic acids and proteins are the information molecules. The sequences in living things often get disrupted—mutations, genetic diseases, and cancers are examples. Essentially the disorder is due to errors in the sequences that represent how the cells of the organism should grow. One way of correcting sequences that have been disarranged is through the use of information, and the source of this information may be our food. The maintenance of life demands a supply of information from the food we eat. The high-level foods for humans are proteins, often animal proteins, which contain just the sort of data needed by the body's regulatory mechanisms to repair damage and maintain sequential information for survival.

The control of sequences demands information and this requires a carefully regulated supply of entropy. This in turn implies that life seeks, not to decrease, but rather to increase its entropy, but it is a gradual and controlled increase, limited by the absorption of just the right amount of negative entropy in the form of food and drugs.

DRUGS

An interesting question is how drugs act on the body. Many chemotherapeutic drugs act by increasing or decreasing the amount of some chemical in the body. However, homeopathic drugs must act in a different way, since they contain so little active constituents and certainly not enough to behave chemically as the chemotherapeutic drugs do.[22] What can be their mode of action? Some investigations were carried out by Jacques Benveniste into the effects of homeopathic medicines. He found that they were effective in concentrations that made it unlikely that there was any chemically active ingredient present. In the case of such a dose, he suggested that the only way in which it could realistically have an effect would be by the information represented by the chemical that is retained by the excipient or medium used to deliver the dose.[23]

During the manufacture of homeopathic medicine the active chemical is not only diluted but the mixture is shaken during dilution to achieve proper mixing. The exact method of mixing is thought to be highly significant in terms of efficacy.[24] What is the implication of this? It is that in the case of such drugs, the mixing process carries a pattern of information from the chemical into the water and that it is the information content of the drug rather than the chemical itself that is effective. Benveniste[25] maintains that this is so and claims to be able to store the information on a floppy disk and transmit it over the Internet.

ENTROPY AND TIME

Entropy in a sense defines time, because entropy always increases with time, because what is measured by entropy is the result of iterative moves, as in a chess game. But the orderliness can be regarded as being the same, with the moves absorbing the orderliness lost by the board. In the case of the chessboard we are apparently adding information with each new move. This makes the board position look more complicated as time goes on. However, if all the moves are written down at the start of the game, for instance, if we are re-playing a game from the newspaper, there is no information added; it is simply interchanged between one form of representation (the written moves) and another (the board position). Thus the moves plus the board position contain constant orderliness at each moment. This could be expressed in terms of information by saying that the information was constant, with the information added by the moves being gained by the board positions. Now these iterative moves do not actually add information to the situation—they simply redistribute it. But if the game is actually being played, is the situation any different? For here the representation is in terms of brain states and their precursors, which are to be found in other events.

In the evolution of the world we have a different situation. Here the moves are not written down or in the brain of a player; they are the outcome of the laws governing particle moves. Since these laws are invariant (if we have found the right laws), how can they add information to the world? It seems to be the process of iteration that adds the information; that is, it is time that adds the information. Information *must* be added, for if it is not, very odd things will happen: chaos theory predicts that, if no new information is added, the world will very rapidly grind to a halt.[26]

LIFE'S LUNCH

As we have seen, entropy is pumped out of living things only by pumping it into the environment. Does the environment become less orderly as a result? Like the effect of using a picnic area to empty your car ashtray, seemingly irresistible behavior for the town dweller, does nature get silted up with life's rubbish? Or, to put it another way, is life having a free lunch?

The answer seems to be no, it isn't. The order that is ratcheted out of the background processes must reduce the total amount of order available. This would assume, however, that there is only a limited amount of order or negative entropy available for us. But that need not be the case: the amount of order latent in the environment may be in effect inexhaustible. In the sea of particles that constitute the world, there may be an infinite amount of order available from what are apparently random processes: so that randomness, rather than being complete disorder, is, in fact, infinite order.

LIFE ITSELF

To take up Schroedinger's question once again: what is life? Life is a process but we do not yet fully understand that process. In our experience, life is always based on a self-replicating molecule whose replication has consequences. The consequences are the material of the study of genetics: the subject of self-replication is central to biology.

What living matter needs to reproduce itself indefinitely is first to be able to copy its own means of reproduction. Once the reproductive system has itself been reproduced, then the system can produce another organism. Since life is a program, the prime requisite for the species to stay alive is for each individual to copy its program—that is, its nucleic acid, whether RNA or DNA—which is its means of reproducing. In the case of double-stranded DNA, this is done by a sort of printing process—making a negative and then another positive—so that you end up with a copy of the original. The bond-ed structure of double-strandedness makes this a natural way of copying, because the negative is already there awaiting separation. But before this stage was reached, there must have been some way in which single-stranded nucleic acids were copied. Another suggestion is that instead of the formation of RNA from scratch, there was an intermediate stage in which protonucleic acid (PNA) became self-replicating (PNA is the analog of the phosphate backbone of RNA), and that the information held in PNA was then transferred to RNA.[27]

Suppose that long molecules of protonucleic acid had been forming and dissolving for a very long time according to a Markov-chain process such as we discussed in chapter 3. Such a chain would become more complex or less complex at random, so the average length of the chains would be short, but some longer ones would form from time to time. One day the chain would have persisted long enough for a molecule to come into existence that had the vital characteristic that it could copy itself. To do this it would have had to embody (we assume by chance) the self-copying program. It would therefore follow that simple nucleic acids should still contain this self-copying program, which should be found if diligently sought for.

During the processes preceding prokaryotic life it may have happened that molecules were arranged in chains containing RNA that was serving no particular expressed function (what are now referred to as introns: these are sequences that do not lead to amino acids) and that one of these chains came to contain the copying program. Once this had happened, further copies would then automatically be made. The length of time it took for such a chain to arise would be the time taken for prokaryotic life to originate.

To verify this suggestion it would be necessary to look for the copying programs in the DNA of some organism. What would the program look like? We can make some guesses as to its probable function and structure and one guess is that it would follow the design of the Turing machine (which is the universal prototype for a computer). It would consist of a self-replicating Turing machine. It is, of course, unlikely that such a program would arise; but much more likely than that all the biological structures we see around us arose by chance, without such a program to generate them.

But how to find such a program? One of the first problems would be to discover its coding and in this decoding process it would be worth considering the fact that there are four kinds of base. There are several possibilities as to how the code is made up from these. One is that all four bases may be used to specify the copy, add, multiply, compare, branch, and other functions necessary for morphology. Another possibility is that the bases may serve as separate organizers in a four-dimensional system; each one containing within its sequence a program in unary code for the replication of the molecule, together with counters for the regulation of the geometry of the body of the organism plus life-span as a fourth dimension.

However, we can perhaps do better than an exhaustive search-and-decode approach. It may be possible to use theory to find the relevant parts of components of a program of this nature and then confirm this informed guess by experiment. Any program contains blocks that perform certain functions. In this case, the blocks consist of copying, counting, and moving the read–write mechanism. However such a program was written, these components must be present and their sequence must be the same, or there must be a constant framework in any program corresponding to the sequence. A machine capable of finding this structure—say a neural network—might be devised and tried out on the likely sequences of DNA to see if such a program exists. It is the sequence of operations that it would look for, regardless of the microcode in which they were implemented.

Life is based on a molecule that (aided by its immediate environment) replicates itself. This on its own would be enough to ensure survival, since if the original were damaged or destroyed, the copies would continue to be able to reproduce themselves, and it would be less and less likely that all copies would be destroyed or damaged as time passed. Such a thing could still happen; for example, all life on earth could be extinguished by an unlucky strike by an asteroid. A self-copying molecule of RNA was the first, powerful evolutionary step and ensured the continuation of life. Prokaryotic life follows this model; but for the eukaryotes a different mechanism, a second copying process, comes into play. This is an internal copying process, which gives rise to the many copies of segments of DNA within the strand itself—the repeat DNA. The program must exist for this copying too, and be found within the same DNA strand. It must exist, ready for inspection, in simple organisms whose genome is fully worked out, because everything that the cell does, it does under the control of DNA. If it exists, then inspection should reveal it. But why make such copies? What is the purpose of it?

If you look at DNA from a mathematical point of view, the question becomes easier to tackle. When the molecule became self-copying, the future of life was assured, but not its development, its onward march toward complexity. This had to await the emergence of structure. One way in which a molecule could determine the shape of the organism that it specified would be by having within it a program for copying segments of itself. These multiple copies will be used by the cell to determine its criteria for growth, gene expression, and organization of sets of interacting genes, or operons.

LIFE AS ENTROPY

Let us try to conclude this discussion by considering what entropy and life have to do with one another. The picture we are building up is of life as a self-organizing system, able to maintain its stability through chaos—chaostability—and allowing a gradually increasing background of information through evolution. The living process is not inherent in one place and time but is linked to the nature of a developing world—a thought that has been voiced by many people and in many ages. Now, however, we can begin to see the detailed description of how such a process functions. Life as such a self-organizing system is a continuum, not simply a late addition to the world. It is inherent in the rules and its beginnings show themselves at all stages and times.

The image to come out of such thinking is of a dynamic process on a grand scale, unfolding through time to produce information pockets. As it unfolds, it generates in certain places and times enough information for living things to emerge; for an organism one must have enough but not too much information; connection between parts but not overconnection.

If there is such a process, and if it is an attractor, then things are drawn to this attractor to produce life in all its parts and aspects. These may be, indeed must be, widely distributed in space and also in time. In many different parts of the world events will have reached the same conjuncture; this accounts for the zeitgeist—the spirit of the time that informs people simultaneously: Darwin and Wallace, Salk and Sabin, Newton and Leibniz. If so, then what we are now thinking and saying here is necessarily being thought and said by someone else somewhere else in the world.

AN ENDNOTE ON RECURSION

The idea of recursion crops up many times in this book. What does it mean? In recursion, a concept is defined in terms of itself. A good example of an implied recursive statement in language is "The American Dream," which may be defined as the right to dream the American Dream![28] Another example in British law is the definition of a racist incident as an incident perceived by its victim as racist.

Traditionally, mathematics has used the word *recursion* to mean the same thing as iteration. However, I have used iteration to mean performing the same operation over and over again, carrying the result forward to the next cycle, while recursion means generating an infinite set of values by making a function one of its own arguments. Since the advent of computers, recursion has come to mean something different from iteration in practical terms.[29]

For example, in working out a factorial we can use the formula:

$$n! = n \times (n-1) \times (n-2) \ldots 3 \times 2 \times 1 \qquad (1)$$

This can be transposed to

$$n! = 1 \times 2 \times 3 \ldots (n-2) \times (n-1) \times n \qquad (2)$$

which gives the same result. However, it can be written in a basically different way:

$$n! = n \times (n-1)! \qquad (3)$$

Definitions (1) and (2) are iterative. They can be followed explicitly, without knowing where they are leading, and they will lead eventually to factorial n. Definition (3) is recursive, and leads directly to the meaning of factorial n. Three things should be noted about it. First, it does not say explicitly what the sign ! means: its meaning is defined by the recursive use of multiplication to link n and $(n-1)!$. This, while it appears to be a circular definition, is not, because the recursive definition of $n!$ defines ! as the operation of multiplying the result of $(n-1)!$ by n.

The second point to notice about recursion is that it is brief. The whole of what we need to know is contained in one statement in as few symbols as possible. The calculation takes roughly the same time to carry out on a computer when programmed as a recursion as when it is programmed as an iteration; it may not be exactly the same, depending on the precise implementation of the routines to compute the answer, but the two calculations are of the same computational order.

The third thing about recursion is that it is theoretically nontemporal. This is a difficult point to convey, because the mathematical notation is not able express logical (as opposed to real-world) time values. But the difference may be summed up like this. Operation (1) is carried out forward in logical time; we start from n and work downward to 1. Operation (2) moves backward in time; we start from 1 and build up to n. Operation (3) is outside time; it takes place in a Platonic world where all mathematical operations are essentially timeless.

AN ENDNOTE ON ORDER

The arguments given above seem to lead to the conclusion that it is difficult, if not impossible, to define the concepts of order, information, and entropy without reference to the system that perceives them; that is, to the human mind. This implies that order and entropy are psychological variables, not physical ones. Can this really be true? If it is, then it means that the world we live in is essentially, in the old-fashioned sense, chaotic; one in which the only perceived order is there as a result of the combined interaction of the perceiver and the thing perceived.

This is at first a very disturbing conclusion, since it removes the feeling of psychological and physical security that we have built upon the reality and solidity of our environment. We seem to take it for granted that the earth under our feet is organized in a way different in kind from that of a liquid or a solid. Yet if what we have been saying is true, then can we accept this? Is not the difference between solid and not-solid one of order? And if we cannot define order without taking psychology into account, is there any basis, either literally or metaphorically, for the belief that we are on solid ground?

It could be argued that order is an objective physical quality, although of a different kind from what has been supposed. There is order in systems, but it has to be defined in a new way: this concept of order is geometrical. When we say a system is orderly, what we are really saying is that it has attractors, and that they form and are dissolved as a result of the evolution of the system through time. A cloud formation seems more orderly than a fog in the sense that we can perceive it as a form, but that is only the sign of our ability to perceive its orderliness; the order is there because the motion of air and water vapor tends to a stable attractor and it is the presence or absence of attractors that prescribes order.

But was there not order in the air and water before the cloud formed? Yes, but it needs a different brain to perceive it, one that does not perhaps yet exist. The separated particles that are to become the cloud were arranged in such a sequence that they would unite in the cloud. A suitably programmed computer could perceive the order in these separated (to us) particles of the cloud-to-be, but we cannot. So the order we see around us is the result of our computer, the brain, acting on the formation of matter into an attractor that has become apparent to our mathematical faculty in exactly the same way as the cloud has become apparent to the eye. The eye and brain see only that order in the world that they are ready to see.

10

THE EFFECTIVENESS OF MATHEMATICS

They do not apprehend how being at variance it agrees with itself: there is a connection working in both directions, as in the bow and the lyre.

(FRAGMENT 51)

UNREASONABLE EFFECTIVENESS

We now approach the heart of a great mystery. If living things are gigantic parallel processing computers, then what is it that they are computing? Well, as we have seen, they are computing their own shapes. But this does not wholly answer the question, for why should those shapes be what they are and no other? The answer given by Darwinian evolutionary theory to this question is that there is no answer, because they are supposed to be the outcome of chance processes. In effect it says that the shapes of life are arbitrary and meaningless. But as we have seen, chance is a vague term, a catch-all for what we do not yet understand. And if what I have been arguing is correct, there is considerable meaning in the shapes of living things, because they are the outcome of the mathematical processes that produced them.

In order to answer the question "What is it that life is computing?" we have to look into the heart of mathematics itself. This might seem an impossible task, no less so because mathematics, for so long unchallengeable in its certainty, has for the past fifty years or more been hanging by a thread over a void of doubt. Our next step will involve us in looking into that void and even attempting the desperate endeavor of snatching back one of the greatest of mankind's intellectual creations from the abyss.

There are two big puzzles about mathematics. One is that it is consistent, that it agrees with itself. The other is that it works at all—in the sense that it corresponds to the real world and what goes on there. The first of these puzzles is a very deep one. For centuries people thought that mathematics was necessarily consistent, although they did not know why. The second puzzle, the correspondence of math to reality, has been called by Eugen Wigner "The unreasonable effectiveness of mathematics."

To the nonmathematician, one of the most puzzling things about mathematics is that it works, and works every time. No matter how often you add the numbers up, they give the same answer and it is always the right one. If the sum does not turn out the same as before, it is because you have made a mistake. Inconsistencies and errors are seemingly due only to the person doing the mathematics, never to mathematics itself.

As a child, I hoped to catch arithmetic out in an inconsistency some day, but I never did. Nine and seven always seemed to make sixteen, no matter in which order you did the counting: arithmetic seemed to work however often you did the sum or however difficult the sum was. Later I discovered that this characteristic of mathematics is a source of wonder not just to children, but to mathematicians as well.

The second, equally great, puzzle is why mathematics works; that is, why is it so effective in giving right answers that apply to the world? The fact that mathematics does work in the real world and not just on paper is quite surprising, because mathematics is concerned with number, which is something abstract, while the world is composed of things, which are concrete. The two seem so different that it is difficult to see why they should fit together so well.

Many theories of the "unreasonable effectiveness" have been suggested. Perhaps the most famous[1] was given by the philosopher Immanuel Kant. He started by observing that there are two kinds of truths: analytic and synthetic. Analytic truths are those that are true because they are a matter of definition: like saying "a horse is an equine quadruped." Synthetic truths are true statements about matters of fact: like saying "It is raining" when it is raining. Mathematics, Kant said, consists of truths that are synthetic rather than analytic: two and two make four because it is found that two things put together with two others always make four countable things in the real world.

This seems to be not so much an answer to the problem as a restatement of it. Why should the addition of two things to two others make four things as described by our counting system? The difficulty is compounded when we get to the more complex areas of physics, in which mathematics seems to be just as effective. When dealing with simple counting this effectiveness seems puzzling, but when we are using mathematical theories to predict probabilities in the quantum world with nearly perfect precision it becomes astonishing.

Kant's answer also seems to contradict the way in which mathematics is constructed: in arithmetic two and two make four because that is how we have defined our counting system. In other words, mathematical truths seem to be just what Kant meant by analytic truths.[2] It makes much more sense, within his scheme, to say that mathematical truths are analytic than it does to say that they are synthetic. But even if we do, that takes us no nearer to explaining the close correspondence between mathematics and reality. We will have to look further than this Kantian dichotomy for an explanation of unreasonable effectiveness.

MATH ECHOES THE WORLD

A more convincing answer to this riddle is that mathematics is true because it echoes, in the formal operation of its rules, the way in which the world actually works. Paul Davies advances this point of view in *The Mind of God*.[3] According to him, mathematics has a structural similarity to natural phenomena. "Our view of the world will obviously be determined in part from the way our brains are structured. For reasons of biological selection we can scarcely guess at, our brains have evolved to recognise and focus on those aspects of nature that display mathematical patterns" (p. 151).

The trouble with this answer is that it is very hard to see why our mental processes have come to resemble those of the world around us. "Reasons of biological selection" no doubt. But how can processes be selected unless they have arisen in the first place? How can these two diverse activities, mentation and survival, be brought together? As Davies confesses later in the same chapter, "there is in fact no direct connection between the laws of physics and the structure of the brain" (p. 156). If so, then it is at first sight very hard to see the link between brain processes and natural processes.

Mathematics (and mathematicians) worked well for many thousands of years with very little in the way of an explanation of its basis; it was simply assumed that mathematics was the foundation of reality. At certain times during this period there was a feeling that counting was the foundation of everything. This is well summed up in Leopold Kronecker's famous saying "God made the integers, man did all the rest." At other times geometry was believed to be the source of mathematical truth; some of the classical Greek philosophers, especially Pythagoras, tended toward this view. Both outlooks have their virtues. The use of counting as a basis seems to make sense of the effectiveness of mathematics in a world where objects remain stable and do not coalesce or divide while you are counting them. Geometrical foundations would make sense in a world that could be shown to be constructed out of geometrical shapes, as the Pythagoreans thought it could. But neither view seems very satisfactory as an explanation for why mathematics works so well and so consistently. The world does not appear to be made of simple geometrical shapes (it would be hard, for example, to find a square on a planet that humankind did not live on). Nor is number an obvious basis for counting real objects. We may be able to accept the basic reality of the positive integers, but this depends on the static nature of the objects being counted. It would not be very much help, for example, in explaining the structure of the sun if we happened to live there instead of on earth. And what is the basis in reality for negative numbers, imaginary numbers,[4] or quaternions[5]? Some cultures even experienced difficulty in believing in the idea of zero. Certainly there is no apparent common-sense link between differential calculus and the nature of matter or radiation.

THE LOGICIAN'S FEAR OF THE CONTRADICTION

One of the biggest difficulties in explaining how mathematics describes the world is that there are times when it does not seem to be able to do so. Logic, which mathematicians at the beginning of the twentieth century saw as the basis of mathematics, had long been plagued with paradoxes, even in classical times. A simple paradox of great antiquity is that of the Cretan Liar: someone from Crete says "All Cretans are liars;" obviously, if they are stating the truth, this statement is itself a lie, which leads to the paradoxical result that, if they are telling the truth, they are lying.

A more advanced paradox is the one proposed by Bertrand Russell, in which we have a card with two sentences, one written on each side: "The sentence on the other side of this card is true" and "The sentence on the other side of this card is false." It takes only a little reflection to see that these two statements contradict one another, and also indirectly contradict themselves. Contradiction is much feared by logicians because they feel that once a contradiction is allowed, any statement can be proved.

How can we escape such self-referential paradoxes? One could say that they are not proper statements, because they do not refer to anything other than themselves, and are therefore empty; such statements are said to be "not well founded." But on what grounds can this be defended? It is normally thought perfectly legitimate to make statements about other statements, for instance, when someone says "You are lying." To prohibit this just when the statements turn out to give rise to a contradiction seems an arbitrary definition of what is a valid statement. And how can we find out that statements give rise to contradictions? Suppose I have a series of statements which say

A: "Statement B is true";
B: "Statement C is false";
C: "Statement D is true";

and so on.

Can all or any of these be accepted as valid or not? If we accept the definition that they must not be self-referential, then we can find out only by working through the whole string and making sure that one of them does not refer to another in such a way as to make it not well founded and thus supposedly empty it of meaning. If we accept as valid only statements that are well-founded, then we will not know whether any of them are legitimate until we have already been working with them for some time! This seems an otiose definition of validity. In any case, there are many situations that occur in real life, for example, in logic puzzles, of just this type, where sets of statements like these seem to be useful and meaningful. How can we condemn these possible worlds of self-referential logic simply because they are self-referential?

PARADOX AND PRINCIPLE

It was not until the nineteenth and twentieth centuries that logicians and philosophers began to try to set down systematic foundations for mathematics, and when they did so it was soon found that paradoxes created immense problems. The climactic and most complete attempt to foundationalize mathematics was made by Whitehead and Russell in *Principia Mathematica*,[6] which attempted to show how mathematics could be derived from a few axioms of logic—the so-called Peano axioms. This attempt seemed at first to succeed but was ultimately undermined by the discovery that there were paradoxes in logic and in mathematics that could not be explained from within the system itself.

These paradoxes arose in the theory of sets. The notion of a set is one of the most general in mathematics. You can have a set of almost any kind of things: a set of books, the set of all blue objects, or the set of all things that are three in number. The last of these is useful in the attempt to lay mathematical foundations: the cardinal number n can then be identified with the set of all things that are n in number and this can be used as the basis of the positive integers (including zero). Once you have constructed definitions of the integers, you can define other kinds of numbers, such as the negative integers and the rational (fractional) numbers. (There are difficulties with real numbers, i.e., those that do not have exact fractional representation, such as π, but I will leave this to one side.) Set theory consisted only of the constructions of logic, so the hope was that mathematics could be founded upon logic, which had, intuitively at least, a basis in reality and common sense.

The difficulty with making set theory the basis of mathematics was that set theory itself turned out also to contain paradoxes. The following is known as Russell's paradox and is about sets that are not members of themselves. Sets may be members of themselves or they may not be members of themselves. For example, the set of all large things is a large set, so it is a member of itself. Another example, the set of all sets is itself a set. Some sets on the other hand are not members of themselves: the set of all books is not a book, the set of all virtues is not a virtue. Now consider the set of all sets that are not members of themselves. Is it a member of itself or not? Some thought will produce the following answer: if it is, then it shouldn't be, and if it isn't, then it should be. If it is, then it shouldn't be, because if it is a member of itself, it should not be a member of the set of all sets that are not members of themselves. On the other hand, if it isn't, then it should be, because if it isn't a member of itself, it should be a member of the set of all sets that are not members of themselves.

Russell made this puzzle concrete by asking: In a town where the barber shaves every man who does not shave himself, who shaves the barber? The usual answer is that, if the barber shaves himself, then he should not, and if he doesn't, then he should! This sort of thing might seem to be merely a tortuous play on words. But if one of the most basic ideas in mathematics—the idea of a set—leads to contradiction, what reliance can we put on the rest of mathematics, if it depends on set theory for its support?

Russell could not at first see any escape from this situation: it appeared to undermine the foundations of logic and hence of a mathematics that was derived from logic, as he was attempting to show. Eventually he suggested that statements might be regarded as not being all of the same type. Statements about *things* would be of one type. Statements about *statements about things* would be of a different type, and an infinity of types would follow, with *statements about statements about statements,* and so on. To talk about *things* as members of a set, which is an object of type 1, is to make a different type of statement from talking about sets, as members of a "super-set" of type 2. Thus, in Russell's theory of types, the question of a set of all sets belonging or not belonging to itself could not arise, because a set is of type 1, while sets of sets are of type 2, and a set of type 2 could not belong to a set of type 1.

FROM PARADOX TO INDECISION

But this was not the worst difficulty for those trying to underpin mathematics. I said earlier that mathematics itself seems never to give contradictory answers—it does not contain inconsistencies. But in the nineteen thirties the Austrian mathematician Kurt Gödel actually demonstrated[7] that it does! He did it in this way: he started with the same system that Russell and Whitehead had used, which was founded on the axioms laid down by the Italian logician Peano. The next step was to assign a unique number (later called the Gödel number) to every distinct arithmetic expression in this system. He then constructed a mathematical expression that had a Gödel number *G* and in logical symbols signified the proposition "the expression with Gödel number *G* is undecidable." This expression is itself undecidable, since the assumption of both its truth and its falsehood lead to contradictions. This was not just a verbal paradox, nor even a sophisticated paradox made with logic symbols, but a seeming inconsistency at the heart of mathematics.[8]

Gödel made the matter even more serious by showing that his result did not apply just to the Peano system. He showed that no system of axioms of this kind could be constructed that would not contain such an inconsistency. In other words, *not all the propositions of a system could be decided within the system itself.* This meant that the foundationalization of mathematics of the kind that Whitehead and Russell had attempted could never be carried out. The unreasonable *in*effectiveness of mathematics had been demonstrated. In view of all these doubts and difficulties, it is a high irony that to the average person, mathematics is the one thing in the world that they would say was certainly true, even if they did not profess to understand its details. In *Nineteen Eighty-four,*[9] Orwell chose to use the statement "Two and two makes four" as the one last-ditch truth that Winston Smith must be made to deny if his will was to be broken and his submission to the Party finally achieved.

ADDING ANOTHER DIMENSION

Can iteration theory help to explain these seemingly paradoxical results? I believe that it can. These seeming paradoxes are the result of viewing the situation in a flat and perspectiveless way. An analogy from visual perception is the Penrose triangle illusion. If we look at it in two dimensions as in figure 10.1a, it appears to be impossible. But if we look at it in three dimensions as in figure 10.1b, we can see that it is possible to construct it. It is three wooden sticks that

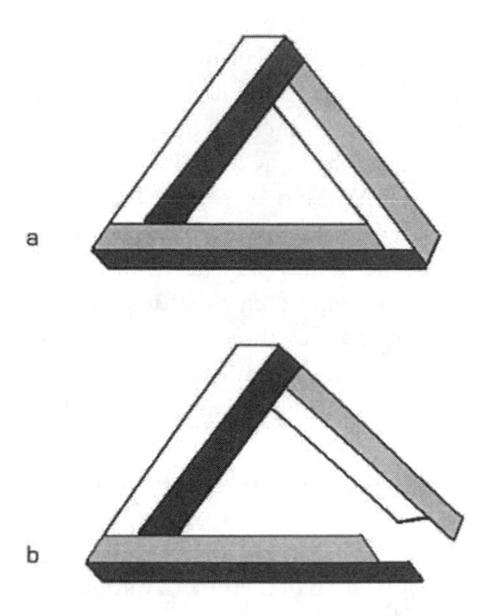

FIGURE 10.1 The Penrose impossible triangle.

really meet at two places, and only appear to meet at the third when seen from one specific viewpoint.

It may be that the paradoxes of logic and mathematics arise in the same way as this seeming paradox in perception. The key observation is that in logic and mathematics, statements are supposed to be true or false in an unvarying and timeless way. If two and two make four, then they will always make four and they will never make any other total. But if the truth of a proposition is not timeless, then its truth can change with time, and this is an accurate model of the paradoxes. Take the case of the card with two sentences on opposite sides:

1. "The sentence on the other side of this card is true" and
2. "The sentence on the other side of this card is false"

If statement 1 is true, then 2 is true also. But 2 asserts that 1 is false, so we have a contradiction, because 1 cannot be both true and false. The straightforward answer, and the one avoided by logicians, is to say that 1 was true but has now become false, rather like the statement "It is raining" becomes true and false according to the state of events that it describes.

Contradiction is not supposed to occur in logic or mathematics, and logicians have what Wittgenstein called "an unreasoning fear of contradiction." One reason why contradiction is prohibited in mathematics and logic is because it is thought that if it occurs, then anything can be validly deduced. Suppose that A is true and false at the same time. From this we can prove the truth of any other proposition B. We start by saying that A is true. Since A is true, we can certainly say A is true or B is true. We can also say that A is false or B is true. From this it follows that B must be true. In this way any proposition can apparently be proved in a system of logic that tolerates contradictions. However, on reflection it can be seen that there is no harm in asserting it to be raining at one time and not at another. Transfer this idea to logic, and we get the axiom that a statement cannot be both true and false at the same time, but it can be true at one time and false at another. The difference between logic and the real world is that in logic, the time is logical time rather than physical time.

To apply this idea to the card paradox, we may say that the statements are alternately true and false—that is to say, they oscillate in truth value. Why do they do so? Because the evaluation of their truth value is an iterative procedure, and their relationship is such that the iteration leads to an unstable truth value. Not all pairs of statements will lead to unstable truth values when iterated. The pair of statements "The sentence on the other side of this card is true" and "The sentence on the other side of this card is true" are fairly obviously both stable under iteration. Even contradictory statements are not necessarily unstable in their truth. Consider the following pair of statements on opposite sides of the card: "The sentence on the other side of this card is false" and "The sentence on the other side of this card is false." These contradict one another, but they do not lead to an unstable truth value. If we assume that the first statement is true, then it remains true and the second remains false.

STABILITY AND CIRCULARITY

Let us see whether we can apply this idea of unstable truth values to the Russell paradox about the set of all sets (if the set of all sets not members of themselves is not a member of itself, then it should be, and if it is a member of itself, then it should not be). The solution by iterated truth functions is to say that the set at one logical time belongs to itself and at another logical time does not belong to itself. (Of course, these times should not be thought of as real-world times; otherwise we get a bizarre picture of propositions becoming true and false at incredible speeds as they re-evaluate themselves. The distinction is the same as that between any theoretical model and the real situation corresponding to it. Logical time may be used in a model of a real process of evaluation carried out either by a human being or by a machine, but it is a distinct concept from physical time, so gigahertz-oscillating truth values could exist only in the hardware of a supercomputer.)

It seems from what has been said that the situation in which these kinds of paradoxes arise is one in which a statement has been defined in a circular way, either directly, as in the Cretan Liar, when it contradicts itself, or indirectly as in the card paradox, when it refers to another statement that contradicts it. The defining feature of these cases is self-reference, direct or indirect. We could go further and say that any situation that includes self-reference of this kind will lead to the iteration of truth values, though not necessarily unstable ones. Gödel's theorem also seems amenable to this approach, since it contains a statement that refers to itself (a statement with Gödel number G that says that a statement with Gödel number G is undecidable). Following this idea, it would seem that the Gödel statement might be true or false alternately each time it is evaluated. This gives rise to the idea that it is also alternately decidable and not decidable. We thus have a proposition that is sometimes true, that some statements in mathematics are sometimes undecidable, but this does not vitiate the whole of the foundations of mathematics; indeed, it puts them on a much firmer footing than before.

To put matters straight regarding the paradoxes, the assumption of an extra dimension of time exorcises the apparent contradictions arising from these old, troublesome paradoxes. It also opens the door to a new foundationalist approach to what mathematics really is. Gödel's theorem has been used in a peculiarly underhanded way by some philosophers who argue that, since a proposition that was undecidable could not be shown to be so within the system in which it had been legitimately defined, the human mind had a capacity for deciding things lacking in any automaton that could ever be constructed. Artificial intelligence could, therefore, never become a reality. But if the proposition turns out to be not always undecidable after all, then this contention falls and we needn't consider it a limitation any more.

If we can remove the set-theoretic paradoxes and circumvent Gödel's theorem, then the prospect for the foundations of mathematics are much brighter. The possibility at least remains open that mathematics is both complete and consistent within a system of varying truth values.

MATH AS ITERATION

Such a philosophy of truth fits well with a more general assumption, that of the iterative nature of mathematics as a whole. All mathematical notation is really just a surrogate for iteration. To get to the number seven, you must add one to itself six times. Each time you carry out this operation, the previous result must be used. Counting is clearly an iterative operation as defined in this way and as such it is one of the most natural operations to be represented mathematically. A consideration of the historical roots of number confirms this view. The origins of numbering probably lie in the tally system, of which evidence can be found as long ago as 32,000 B.P., and which we know was used in later times for recording commercial transactions. A system in use in Iraq in 1,700 B.P. inscribed marks on the outside of clay containers to determine how many units of produce were in transit. One tally would be written as the number 1 is today. If two tallies were to be written one above the other (=), they could be written more quickly by joining them together in one fluid hand movement: 2. Similarly, three marks (≡) became the number 3. Following this line of thought, possible representations of the numbers four, five, six seem clear enough.

To see the origins of our numbers in this fashion is to see their roots in the iterative process. At the time of their invention, the signs that represented integers were a sort of shorthand for the actual number of tallies involved. Rather than doing an operation seven times, they could represent this sequence of repeated operations by the number seven. We can count up to seven, or we can substitute the number seven for the result of the counting operation.[10] From this start it is not hard to progress to the operation of addition. Adding two whole numbers really means taking two strings of ones and joining them. At a slightly more advanced level, we can write the number ninety, or alternatively we can write out the number ten nine times; this would constitute the operation of multiplication.

The use of bases for numbering introduces little in the way of further complication. The use of higher-base arithmetic adds the ability to store the results of previously executed iterations as tables, so that we do not have to repeat the process of iterating every time we want to make a computation. Instead of writing symbols for ever-larger iteratively formed results, a system was introduced of denoting a particular number, called the base (ten in our tradition), by the symbol for one written in a different position; further digits could then be put on to the end in the original position. However, this does not alter the basic nature of the iterative processes involved. Mathematics is as much or as little iterative whether we use base-365 or binary arithmetic. The purpose of using a higher base than one in arithmetic is to shorten the processes involved. In an ideal world we could write every number in unary form, as 111111 . . . 1111, but life is finite and if we want to get an answer more quickly than unary arithmetic will permit us to do, we must adopt higher bases for arithmetic. Lewis Carroll's White Queen asked, "What's one and one and one and one and one and one and one and one and one and one?" "I don't know," said Alice, "I lost count," to which the Red Queen retorted, "She can't do addition." No one can do very much unary arithmetic, although nature can. The use of a higher-base system enables us to predict economically the outcome of the iterative process.

It is clear that many constructs in mathematics can be analyzed in the same way, even though they are more complicated. For example, we can represent an integral as the result of adding many slices of a quantity; these are typically slices of space or time.

"PROBLEMS TO FIND": PARTIAL EVALUATION

I am suggesting that mathematics is the application of a shorthand notation representing iterative operations to situations where actual iteration is involved and therefore it works because it shares the same structure as the events it represents. Two main kinds of evidence support the idea of mathematics as iteration. The first is that it is in situations where there is physical iteration that mathematical analysis works. The revolution of bodies under the attraction of gravity, the collecting together of like objects, the combination or permutation of events, the rotation of planes to form solids—it is in situations like these that mathematics has had most success. The second, and converse, piece of evidence is that it is in situations where continuity breaks down that mathematical treatment is most difficult or impossible to apply, for example, during turbulence at boundaries, or in relations of disparate phenomena. Under these circumstances either there is no relation between the repetitions during one cycle and the next, or, more fundamentally, the prerequisite repetition of operations seems to be altogether lacking.

Algebra at first seems to pose problems for an iterative theory because no real counting is apparently going on. However, there is implicit counting: algebra can be seen as a type of Polya's[11] "problems to find." The question that algebra asks is: What is the iteration that will give the specified result? This is represented as an unknown. If we are asked what is the whole number that, when added to half of itself, will make nine, we can try out various iterations until we get the right answer. We are then asking: for what x is the iteration represented by $x + x \div 2 = 9$ true?

It is perhaps at first a little hard to see this as a genuine iteration, but it may help to consider the problem in the following way. The type of computer program known as a *partial evaluator* reads in a program together with some partial data and produces a new program. The new program, when it is executed, reads in the remainder of the data and will produce the result. This is what is going on in a computer when it is asked to do algebra, and this is the actual iterative process that is represented by the algebraic formula. This is also, I suggest, what we do when we do algebra. The human mathematician has a program for iteration (partial evaluator); takes in the equation (partial data) and produces a new program; tries values in it (remainder of data); continues the process if the equation is not satisfied; then terminates the process if it is satisfied.[12]

What about other types of mathematical objects? It seems that the world of geometry is very remote from the operations of iteration. But geometry can be done without drawing pictures: it can be reduced to stating propositions and manipulating them to produce proofs. It would be easy to give many examples of this. In the late nineteenth century the propositions of Euclid were reduced to logical rather than geometrical terms: geometrical theorems emerged from algebraic formulas and the visualization of the form became an aid rather than the primary medium. In the behavior of the chaotic logistic function, the bifurcations can be seen to emerge either from the formula (the Feigenbaum ratio) or from its geometry when plotted as a graph (see chapter 4).

COMPUTING THE UNCOMPUTABLE

We can now propose an answer to the question of why and how mathematics "works." It works because the iterative processes of the world correspond to the iterative processes of mathematics. Sometimes these iterations are explicit, as in the case of counting, multiplying, or forecasting the orbits of variables in phase space. Sometimes they are implicit, as when we use differential operators to model the gradient of an electromagnetic field. In the first type of case the correspondence is obvious; in the second we have to make a roundabout approach by summing tiny slices of space/time—infinitesimals. However, in both cases the iterative process is still there.

Much that is effective in mathematics can be done by computers, which are new devices, introduced only in the twentieth century. What is the relation between the computer and the brain? Without computers I suggest, mathematics works by setting the brain's state into a correspondence with the results of an iteration: computers work by setting the brain's state into a correspondence with iterations carried out in a computer. Computers give us a handy way of seeing what our brains are too slow to see. They can also show us something like Gödel's theorem unfolding before our eyes. An ancient problem (in computer terms) is how to tell whether a calculation given to a computer can terminate or not. (This halting problem actually preceded the computer itself, because it is related to calculation, whether done by a machine or with pencil and paper.) In 1937 the British mathematician Alan Turing wrote a paper that was, in its way, as devastating as Gödel's earlier in that decade. In it he showed that the halting problem is undecidable: we cannot tell in advance whether a number is computable or not. This left the implication that there are uncomputable numbers—numbers that a computer could never print out (though we cannot of course name them)!

Turing's argument essentially consisted of "defining" a number that was not only uncomputable, but consisted of an infinite number of uncomputable digits. The validity of this argument really depends upon whether a definition is an adequate guarantee of existence. I can define a number that is the next even prime[13] higher than 2, but it does not exist according to any known criterion. Let us consider the largest number that can be represented in the universe, *L*. No larger number can exist.[14] This contradicts the well known proof that there is no largest integer; according to this view there is a largest integer, though its value may change over time. So it may be that Turing was wrong, as Gödel was wrong, or rather that they are right only within a timeless mathematics.

OBJECTIONS TO TIME IN MATH

The point of view I have been expressing will seem repugnant to mathematicians who see the subject in a timeless, realist way; these are almost certainly in the majority. For example, in *The Emperor's New Mind*, Roger Penrose[15] takes an explicitly realist stance: according to his outlook "mathematical objects have a timeless existence of their own, not dependent on human society or on particular physical objects." John Barrow[16] in *Pi in the Sky* takes a similar line. But to be viewed as iteration, numbers most certainly depend on physical objects, because without some kind of computer—human, natural, animal, or electronic—no numbers could exist; and they depend on time, because if they are yet to be computed or still being computed, they do not yet exist. The psychological underpinning for this would be that the nature of mathematics corresponds to the iterative processes in the brain. For the most part these processes are unobserved; but occasionally we have access to them and then we have the perception of a mathematical truth. If we can perceive them, then we can denote them by a suitable symbolism. However, what we have really experienced is the iterative process itself, which, as is well known, can proceed unobserved and often in undenotable form in *idiots savant.*

There are various objections that could be made to this way forward. The simplest (and perhaps easiest to deal with) is the objection that it makes nonsense of mathematics. Are we not saying that two and two do not always make four, when we know, as surely as Winston Smith knew when he was in his right mind, that they do?

Such a view would indeed be incompatible with common sense and I am not saying any such thing. The question of what is always true in mathematics (as two plus two makes four is) depends on what happens when we iterate the formulas of the statement. Some mathematical statements will be stable under iteration, and those of everyday arithmetic, such as two and two make four, are stable statements, having a truth value that does not change. But other statements, particularly those involving self-referential definition, are unstable. It is these that require the theory of logical time to help us out of the paradoxes that will otherwise threaten to engulf mathematics. So two and two will always and forever continue to make four.

Another objection is that the path of logical time is one that need not be followed. We could persist with timeless truth values in logic and timeless mathematical statements, and be content to accept the situation of the undecidability of some mathematical and logical statements, the latter incidentally of a very peculiar kind. Changes and advances in mathematical notation have been continuous over many centuries: at one time people were reluctant to accept the use of imaginary values in mathematics because they were held to be meaningless or unreal. This seems quite absurd to a modern mathematician. A meaning was found for imaginary quantities and they are indispensable, not just in pure mathematics but also in many applied areas. But more important is the fact that they were soon accepted as a theoretical advance, and this has been typical of mathematical advances in notation of this kind. What happens each time a notational advance is introduced is that the range of assumptions is progressively widened to allow new constructions to enter. At first such ideas seem strange and exotic, but soon they are assimilated and taken for granted.

TIMELESS TRUTHS?

There is a deeper objection, which is not often articulated and is basically a philosophical one. Ever since Plato, there has been a prejudice toward regarding the truths of mathematics and logic as essentially timeless. Plato postulated the existence of forms that existed outside the world of sensation; these forms were the idealized versions of the things we perceive and included the objects of the mind. Mathematical truths as expressed by human beings will also have their counterparts in these forms, and since the forms are outside time, it seems to follow that the mathematical statements made in this world must also be outside time. Of course, we can't escape from time in the real world; it takes some time to write and read the formulas in mathematics. But the way around this was to suppose that a formula had a once-for-all truth value, and that if sets of formulas were used, as they are in proving theorems for instance, then all the formulas were true together and simultaneously. In such a system it is not possible to allow the truth value of a formula to vary, because once this is done, contradictions will arise, with all the disastrous consequences this was thought to allow.

To some extent this objection has already been answered: such timeless truths can only exist if they correspond to stable iterated values. But I will consider another part of the answer in a moment: unstable truth values, unstable when iterated, that is, can give rise to a different kind of stability, that of the production of forms, created over time; and the forms may be quite concrete and perceptible and are an important step toward understanding biological forms and their significance. Their existence is also a confirmation of the Platonic idea, as we shall see in chapter 12.

TRUTH VALUES AS MORAL VALUES

The final objection to consider is a moral one. Truth is usually assumed to be a sacred value, on a level in many people's minds with honesty or goodness—a kind of virtue in fact, having a moral quality. It comes as something of a shock to a maturing child to realize that truth in matters of fact is often treated not as absolute but as relative, and is frequently conditioned by viewpoint, motivation, or even by the mere convenience of adults. However, for most people there is still a belief in a special kind of truth—mathematical truth. We all know that two and two make four. This kind of truth exists in a world, as Russell himself said, that lies outside our human concerns, and this kind of absolute truth still survives, even in the disillusioned heart.

Probably this attitude toward truth has made people reluctant to question some of the characteristics of true propositions that are not necessarily essential to the idea of truth itself. The idea of truth may be extended beyond that of timeless truth, and I have tried to show how this might be done. It might at this point be appropriate to ask whether the proposed solution, by introducing the idea of variable truth values, does not tend to weaken the moral status of truth. As I see it, the answer is definitely not. The reason for the fluctuations in people's beliefs, feelings, and statements about them probably lies deep in the iterative structure of living things themselves. It is in the nature of living things to iterate, to undergo cycles of change in regular yet subtly changing patterns. Far from the truth being weakened by adopting a logic that admits of changing truth values, we may hope that our under-

FIGURE 10.2 The Mar and Grim "Dualist" escape map.
By kind permission of Professor Patrick Grim.

standing of truth and our use of it will be made clearer by an acknowledgement of its real nature.

We have already discussed in chapter 5 the shapes of living things and their relationship to iterated mathematical formulas. Now we can begin to make a link between these shapes and those produced by the iteration of formulas of a particular type—those representing logical statements. These statements are, as it turns out, the kind that oscillate in truth value, leading to paradox. It seems reasonable to extend our findings to suggest that the shapes of living things are the expressions of such logical statements. If so, then the very structure of life is an expression of iterations that represent propositions. In 1990 Gary Mar and Patrick Grim wrote a paper[17] that shows vividly what happens when you iterate statements with varying truth values. They assumed that truth values can take any value between zero and one, not simply the two integers representing truth and falsity.[18] The resulting shapes of the escape diagram[19] are interesting. One is shown in figure 10.2, which represents a paradox called "The Dualist."[20] The organic appearance of such diagrams is remarkable. It seems an obvious next step to suggest that these organic shapes are related to the iteration of formulas carried out by the genetic computer postulated in chapter 7.

TRUTH AND BEAUTY

These shapes are the expression of iterated formulas that could find no other way of being expressed in the world. Here, geometry and the propositional forms of mathematical representation are shown to be linked in a close and inextricable way. It is only through the iteration of these formulas that such shapes could come to exist. The natural world is as it is because of computation, and the form of that computation is expressive of the structure of the living things themselves. Could we go further and say that this casts a light on the subject of the innateness of language ability in living things? People are able to speak and communicate in words because of the special nature of their vocal apparatus, in particular, the ability of the lips, tongue, and teeth to form consonant sounds and combinations of sounds. Without the consonants words would lack the necessary qualities of memorable distinguishable stimuli: an experiment that tried using a purely vowel-based language failed. We need consonants for linguistic performance. But linguistic capacity is another thing. A deaf and dumb person may have the capacity for language and yet be unable to use it. A child probably has the capacity for language from before the time of birth. Chimpanzees can communicate using sign language tokens, although the communication is not by means of sounds. What then is linguistic capacity? It is the capacity for the representation of patterns that constitute propositions, and for this a structure is necessary in the brain. But we have seen that life is the shape it is because of the iteration of a universal formula, and the shape is similar to the one we get when we iterate linguistic propositions in mathematical symbolic form. It may be reasonable to posit a basis for this capacity in life itself. What differs between the species then, is the degree to which they can express it.

The second remarkable feature of Mar and Grim's diagram is that it is specifically a paradox that is being iterated. It is as though at least some living forms were founded on contradiction. And again all the moral, social, and personal experience of humankind bears this out. Who does not feel themselves at some times to be a seething mass of contradictory impulses, thoughts, and feelings? As Walt Whitman said,[21] "I contradict myself? Very well, then, I contradict myself, I am large, I contain multitudes." Is this not the universal experience of those wrestling with dilemmas, compulsions, obsessions, moral strivings, and ambitions; in short, of everyone who is alive?

All the esthetic experience of humankind reinforces this point of view. The power of organic shapes in art, esthetics, and all forms of sensual experience is very great and this power has influenced writers, artists, and poets. Such is the power of these shapes to command the human imagination that it is impossible to credit unless their formula is indeed written into our very cells. The organic, dendritic form is universal in culture and myth as well as in biology. From neuron to family tree, and from the Mandelbrot set to Yggdrasil, the Nordic World tree, the shape is found everywhere, the product of changing, often conflicting, values. Would it be surprising if we found it in language and in logic too? If these shapes are so powerful that they dominate our symbolism and feeling, then surely they are involved with the iteration of paradox. Life is contradiction taken to seemingly boundless lengths and it assumes many guises: the fluctuation of desire; the dynamic feeling and emotion; the personal conflicts that can construct the contradictions of human perception as shown in drama, painting, music, and morality.

Whether it is in our own lives or in reading about the lives of others, in the experience of our own or of someone else's work, we know that human existence is a series of contradictions, setbacks, revulsions of feeling, and twists and turns of internal as well as external fortune. It is these continual contradictions, this never-resting search, these near-paradoxes, that provide the dynamic of chaos as an essential feature of life and life-like systems. Life, we might say, is founded on contradiction and paradox, and yet perhaps the greatest paradox of all is that without paradox and contradiction there would be no life.

11

LIFE AND CONFLICT

To god all things are beautiful and good and just, but men have supposed some things to be unjust, others just.[1] (FRAGMENT 102)

PARADOX AND CONTRADICTION

If what has been said about the nature of mathematics and its relation to life is correct, then an entirely new view becomes possible both of the world and of life as part of the world. In particular, our view of living processes will change profoundly, and we will come to see life not as an accidental excrescence or something thrown up in a unique isolated backwater, but as the logical expression—one might say potentially the highest expression—of the processes that govern the world. In Heraclitus' terms, life may be seen as the logos in its highest form, and human life, in our world at least, as the highest expression of the logos.

To see how this is so, I want to return first to mathematics and its relation to the world and more especially its relation to biology. As we saw in chapter 10, mathematics and the world can run in parallel because they follow the same iterative processes. The iterative nature of the world corresponds to the iterative processes of mathematics; that is why mathematics is, in Wigner's description, so "unreasonably effective." In the world and in mathematics, any degree of complexity can be produced from initial simplicity by iteration. In mathematics, the most complex objects that can be produced are fractals, which are the outcome of iterated formulas, their complexity resulting from the iterative process. Iteration can produce form, the degree of complexity of which increases as the iteration number increases. In the real world also, forms are produced by iteration, the complexity increasing over time, while time is itself an iterative phenomenon.

Behind this parallelism of nature and mathematics, as we saw, lies the basic resemblance of two apparently very different strands of work, one carried out by Turing and the other by Gödel. They both showed that within any rule-bound system there are undecidables, whose value cannot be determined within the system itself. Gödel showed this in mathematics; Turing showed it for computable numbers. It is interesting to see the similarities and the differences between their work, a comparison that was recognized by Turing in his 1936 paper.[2] For Gödel the undecidables were the truth values of certain special, self-referential formulas that could be constructed in any mathematical set of axioms. For Turing the undecidable was the ability of a machine to compute certain numbers that he showed to be equivalent to the nonhalting behavior of a Turing machine. Where Turing and Gödel are united is in showing the limitations of any system of symbolic representation, whether it is one conceived by the human mind (Gödel) or embodied in an idealized computer (Turing). Where they differ is that Gödel was concerned with supposedly timeless things—mathematical formulas—while Turing was concerned with things occurring in time—numbers written to a computer tape.

DECIDABILITY AND TIME

In the last chapter I suggested that Gödel's findings on undecidability can be seen in terms of the limitations that have been imposed on our thinking by a realist philosophy of mathematics. The contradiction generated by Gödel, and the resulting theorem showing some formulas to be undecidable, only holds good in a system where mathematics excludes time. If time is incorporated into our notation, we can avoid the paradoxes that have paralyzed thinking about logic and mathematics. Seen in this way, the formulas Gödel was considering are not timeless after all; they are simply appositions of truth values, iterated in logical time. In undergoing this iteration they express the same principle as do Turing's programs for uncomputable numbers. In both cases, what is being computed is a *series* of values rather than one, fixed answer.

If we extend mathematical notation in this way, the distinction between iteration and the mathematical operation of recursion becomes clear. There has long been a custom in mathematics of regarding iteration and recursion as the same thing. If you look up iteration in a mathematical dictionary, you will find something like "another word for recursion." Seen from the standpoint of a philosophy of timeless truth, this is perfectly understandable, but when logical time is considered, a clear distinction between the two processes can be seen.[3] If instead of thinking about mathematical expressions, which are essentially static and lacking in a time dimension, we turn to computing languages, which can express not just formulas but their relationship as events in time, then the difference becomes clear.

An iteration is cast differently from a recursion in computer language and it is implemented differently when it is executed by the computer. Iteration involves repeatedly evaluating an expression in terms of its previous value until a terminating condition or limit value is reached. Recursion defines an expression in terms of itself. The two forms are the same in terms of the final computed result, but in another way they are not equivalent. Both involve repetitive operations; both can be done by means of a digital computer program or by humans following an algorithm, but what a difference between them! The first implies a sequence of steps in time; the second implies all times taken together. Briefly stated, though this answer requires a little amplification, recursion is the inverse of iteration; it is an iteration turned inside out. An iteration can always be stated recursively and a recursion can always be stated iteratively. In the case of iteration, you know what you are doing as you go along; in the case of recursion, you only find out at the end, when it is almost too late.

Seen in this way, recursion is not only the inverse of iteration; it may be thought of as the definition of a function without regard to time. As George Spencer-Brown put it, "A way of picturing counting is to consider it as the contrary of remembering."[4] Iteration takes time, recursion inverts the time sequence and turns it, not just backward, but inside out. The temporal sequence of operations in recursion is the inverse of those in iteration, so another way of thinking about the difference between recursion and iteration is in relation to time. If we look at it from the point of view of some divine programmer external to our world, for whom time can be seen as a whole, in a single moment, we could say that God writes recursive formulas for the world, while we are condemned to live through their consequences as iteration. From the human standpoint, life is just one damned thing after another; from the divine standpoint, it is just one damned thing *before* another.

PUTTING ONE THING BEFORE ANOTHER

If we want a result by simple iteration, we might write a program that worked out successive values of the formula. In computing, a program that invokes a subprogram is said to *call* it; for example, a program might call a subroutine to calculate the square root of a number. If we define a function recursively, we can write it as a program that calls itself—that is, one that is self-referential. This recursive call, whereby a program invokes itself, will generate a series of results. However, such a recursive definition is executed very differently from an iterative definition yielding exactly the same set of values. To the computer scientist, while iteration and recursion may be equivalent in terms of the values they produce, they are very different in the way in which they are executed by the computer.[5]

To shift viewpoint slightly, we could look at these two processes, iteration and recursion, from the point of view of pure logic. It is then apparent that they are in fact inverses of one another. A recursive definition generates all its values instantaneously and by implication, without any elapsed logical time, while iteration implies a series of steps within logical time. A good example is that of laying bricks, which we stated in chapter 1. As we saw then, all bricklaying is iterative. But now we could consider two definitions of a structure of bricks, say, a pile of bricks one on top of another. If we define a pile of bricks as one brick placed on the ground, and another placed on top of the first, and another placed on top of the second, and so on, we are generating a series of steps that will eventually lead to the finished pile, the height of which will be defined by the iteration limit. But we may quite reasonably define a pile of bricks as "A pile of bricks is a brick resting on a pile of bricks." Here we have done something very different. The definition is now a recursive one, and the recursion bypasses the iterative steps. The concept of a pile of bricks, instead of being generated over time, is defined for all the possible numbers of bricks and in one operation. It would not be right to say that the operation takes *no* time; rather, it takes all times together—it collapses time. Of course, we can't lay bricks recursively in the real world but we can in a computer, or at least we can simulate doing so. And we can do it in one of two ways, either iteratively or recursively.[6]

RECURSION AND LIFE

Systems scientists study processes that seem to be defined in terms of themselves: these are the so-called autopoietic (self-making) processes[7] occurring in organisms, for example, those of cell function and self-regulation. Psychologists study processes concerned with human consciousness and will. One of the unique features of consciousness is that we know we are aware. We also observe the results of our own actions, and in this way we learn. There is an analogy here between recursion and consciousness: we know about our conscious processes (consciousness is recursive) while we do not usually know about our vegetative ones (the heart beating, the digestive system working), so we can say that consciousness is a recursive phenomenon while reflexes are not. It is similar when we observe ourselves in moments of self-awareness: our minds then appear to us to be recursive. From the point of view of another mind we simply appear iterative, of course, since we can only observe the behavior of others and not the working of their minds.

We also saw in the last chapter how both formulas whose truth cannot be decided and machines that do not halt may be equivalent processes for producing iterative patterns. Those patterns of iterated activity, either of truth values or of machine states, represent, not simple undecidability, but the forms created by the process of trying to arrive at a decision. In attempting to decide the truth value of certain formulas, we go through a series of transformations that are capable of producing fractal geometrical forms. We might say that these patterns are the solutions to undecidable sets of iterated formulas. We have shown in chapter 7 how the iteration of very simple formulas can give rise to organic shapes and how it becomes possible to approach biological form and the process of its development and evolution in terms of iterated formulas. Various theories were considered as to how this process might be embodied in biological mechanisms. The preferred explanation was given in terms of an intracellular Turing machine, but this may be only one of the possible mechanisms. What is fundamental is the iterative process itself, because it is this that gives rise to the geometry of biological forms. As we have seen, it is impossible to think of any other process that would yield the necessary degree of complexity in so few operations. Random variations in genetic material are by themselves hopelessly unproductive. There must be some principle directing the pattern-formation processes of biology.

So living forms themselves could be regarded as the representation of solutions to otherwise undecidable formulas. They are the embodiment, we might say, of the processes of iteration of those undecidable formulas. But what can be expressed as an iteration can also be expressed as a recursion, so we could also say that living forms are the real-world embodiment of recursive, self-referential formulas. The recursive formula embodied by an organism could be thought of as an existence-proof of that organism, or equivalently, the organism could be understood as a demonstration of the recursive formula. This has profound implications for our understanding of the human condition.

CONFLICT IN LIFE AND ART

The view that life is essentially based on conflict is an ancient one, and it is found in many of the old religions of wisdom. Sometimes this idea of conflict was embodied in the form of opposing forces that created or formed the world, for example, the Yin and Yang of Taoism. Sometimes it was personified in the form of gods doing battle with one another, as a parable, or even an explanation, of the perpetual conflicts of life.

As we saw in chapter 4, chaos theory tends to a view of things in which dynamical forces shape the world. Opposing forces are essential in any nondissipative dynamical system, and since life is such a system, it must be true there, too, that opposing forces shape it. This fundamental fact of the ineradicable nature of opposing forces appears in human terms as conflict, and we can see the precursor of human conflict in the dynamical situation.

Without conflict, life would be impossible. Opposing forces regulate our breathing, the beating of our hearts, the balance of our sympathetic and parasympathetic nervous systems, and the electrical activity of the brain. Physiological measures vary continually and, as we saw in chapter 5, the more complex the pattern of that variation the healthier the organism that embodies them. As it is in the organism, so it is in social relations. Forces tug us in different directions, loyalties and principles conflict, and out of this clash comes the mix of human experience. The same opposition of forces may be spoken of in philosophy as the dialectical process of thinking or the contradictory forces of historical conflict.

If conflict is an inescapable feature of life—if, indeed, it is the very nature of life—then we must recognize and come to terms with this fact. It is essential to dramatic art, for example, where it appears as conflict between characters, ideas, or courses of action. This finds its most powerful expression in tragedy (though it is also present, in a very different way, in farce). It is interesting to see the way that the nature of tragedy has changed over the centuries, showing a progressively broadening insight into the true source of humankind's discontents. To the Greeks, tragedy was often about conflicting principles of conduct, and the source of the conflict was outside the personalities of the protagonists—often two rather than one. The tragedy of King Oedipus is about the pressure of external circumstance leading to a breaking of the social taboos against incest and patricide: the forces of destruction come from outside the man Oedipus (although there is a hint of weakness in his failure to recognize the nature of his unwitting acts). The clash between Antigone and Creon is essentially a conflict of opposing principles: she wants to bury her dead brother for love of his memory; Creon wants his body left to lie unburied as punishment for the crime of rebellion against the state. This conflict of personal feelings and duty to society is a timeless theme. But while later this conflict might appear in a single individual, in Greek tragedy each of these opposing forces was embodied in a separate character. Antigone is not in two minds about wanting to bury her dead brother, nor is Creon hesitant about the necessity of the claims of the state. It is instructive to contrast this treatment of conflicting forces with a more modern tragedy—Hamlet, say, or Othello. In both these plays the central character himself is torn apart by internal conflict as well as being under attack from the external forces at work against him. Hamlet wants to kill his uncle to avenge his father's death, but he is restrained from doing so by a complex of internal emotions involving his feelings toward his parents. Othello is torn between his jealousy of Desdemona and his love for her: he exclaims, in contemplating her murder "The pity of it, Iago!" And it is in this internal struggle, and his failure to resolve it, that much of the intensity of the tragedy consists.

This shift over a two thousand year period from the total externalization of the source of conflict to a partial internalization of it reflects a changing consciousness of the origins of the conflict that is inseparable from man's existence and that tragedy represents in its highest and most clearly visible form. To the Greeks the source of conflict was essentially external; to the Elizabethans it was partly external but also internal. In more modern times tragedies have emerged that consist almost entirely of an internal dialogue in the mind of one person—of Raskolnikov in *Crime and Punishment* or of the protagonist in Camus' *L'Étranger*—something quite unthinkable in an earlier age.

The arts differ from one another radically in nature, and while the emphasis of all serious drama throughout history has been dynamic—the interplay of opposing forces—the nature of music and the visual arts also depend upon conflict as a necessary part of their creative tension, even though the nature of the conflict is not always as explicit as in the theater. These art forms also have had a very different character in different periods. Art in the classical period reflected a synthesis of forces rather than a struggle between them, however necessary the struggle might have been in producing the final state. A crucifixion was essentially the depiction of a finished event rather than the portrayal of a conflict. The Pietà, the taking down from the cross, was a favorite theme: the agony is at an end, for Christ if not yet for mankind, so the feeling produced in the onlooker is one of passive acceptance rather than anger.

Similarly in classical music, the works of Bach represent the reconciliation of thoughts and feelings, any conflict being sublimated, temporary, or the prelude to a resolution. Later however, in its romantic period, the nature of music changed. The music of Beethoven's middle period was essentially the ongoing clash of opposing forces; his symphonies, and perhaps even more his earlier string quartets, consist of a titanic battle between themes. Romantic painting also moved to more dynamic forms of representation with El Greco, Goya, Van Gogh, the expressionists, and later still, the modernists. As the novel develops we see a transition from the more descriptive approach of early works, through the incorporation of human feelings into nature (the pathetic fallacy), to romantic novelists such as Thomas Hardy, who, referring to *Jude the Obscure,* said that a necessary accompaniment to the catastrophe of a tragedy was "the clash of breaking commandments."[8]

THE ATTRIBUTION OF CONFLICT

There is in the West a clear progression in the way in which themes of conflict have been treated in the arts and in the societies that valued those arts. And this shift has had largely to do with the way in which people have perceived the source or locus of the conflict that they have always acutely felt. This progression may be divided into three or possibly four phases. In the first phase, the conflict was attributed to a source outside the human mind, with the implication of the possible existence of an Arcadia, a perfect state, in which conflict could be eliminated. Second, in more recent times, the conflict was seen as being partly internal, but still with a longing for the Arcadian state in which, in the absence of conflict-inducing forces outside the individual (Shakespeare's Forest of Arden, for example, in which the only enemies to be found were winter and rough weather), happiness would be possible. The third phase followed, with the philosophies of Hegel, Marx, and Freud, as it began to dawn that conflict is necessarily internal and inevitable. Perhaps a fourth phase awaits: the realization that conflict is not only inevitable but essential; that life is, of its very nature, generated by conflict and contradiction, and that only in the acceptance of this fact is there any hope of release from it.

The traditional releases from conflict lay in action, in drugs, or in the cultivation of wisdom. In the East, the highest form of wisdom came to be embodied in religions of mysticism, such as Hinduism, Buddhism, or Taoism, in which release was found through mystical contemplation, which alone could set the mind free by putting it into harmony with the surrounding world.[9] This outlook saw that the source of conflict was in man's own nature as well as in the outside world and that only some form of mystical contemplation would enable him to accommodate to this conflict.

The alternative to this religion of wisdom, represented by the Judeo-Christian outlook, took the view that the source of conflict was ultimately external. The force of good—which was God—lay in an externally located reality. This reality was not just spiritual, but physical—the kingdom of heaven is to be established on earth, and it is in my flesh that I shall see God. Similarly, the source of evil—personified as a fallen angel—was also a lively external reality. The human heart was a battleground on which these two powers fought. But there was also a larger cosmic battleground that predated the creation of humans and possibly that of the world itself. The conflict in the human heart was thus merely a by-play, and the source of discontent was located elsewhere: salvation could only be obtained by throwing in one's lot with the better (or, in some heretical views, possibly simply the stronger) party. The objective nature of evil as well as good are essential to this religious view. These mythical forces are to be believed in literally, and it is them that we must blame as well as ourselves.

As a model of reality this outlook was deficient in a number of ways. To start with, it leaves the ultimate nature of good and evil unexplained. Why should the worship of God be preferable to the worship of Satan? What was the origin of evil in the shape of the fallen angel? How and why did God create him? Or if he was not created by God, did he have an equal status in the cosmos, as the Manicheans believed? Insofar as people were led to believe in the objective reality of the forces of good and evil outside themselves, they became warriors in the name of invisible powers, an outlook that has led to some of the grossest atrocities in the history of the human race. The price of beliefs must be paid, and in this case it was paid by the victims of the medieval Christian churches. In Spain processions are still held in which effigies of the devil made out of paper are blown apart by fireworks concealed inside them. This outlook, founded on a realized, actualized myth of the triumph of good over evil, might properly be termed the Religion of Unwisdom.

In contemporary Britain belief in the external and objective reality of a god is waning. Vicars of the Church of England now profess their disbelief, joining the company of their bishops, who have apparently long done so. God is believed in now, not as an external reality, but as some force present in the human heart, offering a hope of the ultimate harmony of all people. It is perhaps significant that, in England at least, belief in the external reality of a god has not long survived abandonment by the Church of the belief in the external reality of the devil. Without the one, the other lacks conviction, because the disharmony of the world becomes inexplicable. As long as God was waging holy war against the Devil, our duty to Him was clear; but without that great adversary, God is reduced from the role of warrior king to that of a purblind social worker; ineffective in the prevention of evil, since so many evils take place before He can, like Superman, rush to prevent them. Before His death, the evil could be blamed on the Evil One, but no longer. The twin beliefs in good and evil are necessary counterparts in this exegesis of conflict, and without one the other collapses for lack of support.

THE ACCEPTANCE OF CONFLICT

In the light of this analysis, what view should we take of the conflicts and contradictions that occur so frequently in our lives? If these are inseparable, not just from the contingencies of life on this planet but from any organic life form, how can we ameliorate or tolerate our suffering? Perhaps the most important, though controversial, modern contribution has been Freud's. He offered a diagnosis but apparently no solution, although many variations on his theme have been tried. The source of conflict, according to Freud, is our conflicted nature, split between rival parts of the personality: the instinctive (Id), the social (Superego), and the conscious mediational process (Ego). To become conscious of this conflict was to be freed of its ill effects in distorting our view of reality, though suffering in the form of toleration of a (now realized) conflict must still continue, and perhaps even increase.

Jung's view differed in content though not in type. Universal and unifying spiritual needs were added to the purely selfish or merely contingent social prohibitions of Freud. However, no relief, even of the symptoms supposedly based on our mistaken understanding, can be shown to be ultimately obtained from therapies based on either Freud's or Jung's doctrine, and this has caused many people to reject their analyses as false. This is a mistake, since there is no necessary logical connection between diagnosis and cure: but many people are left feeling they would do better to look elsewhere for an explanation, as well as a solution, of their woes.

There are important insights here that must be combined with our current knowledge of biological and physiological processes to build up a more complete picture. Humans are not short of sources of conflict, and at least one is that they have a brain with three different functionalities. The forebrain, identifiable mainly with the cortex or rind of the two large hemispheres, is concerned with higher faculties, though the detail of how it functions are still sketchy. The second division is the midbrain, containing the limbic system, which is concerned with emotional functioning, though it is also apparently crucial in memory. The hindbrain, containing the reticular formation and the small hemispheres of the cerebellum, carries out automatic unconscious processes of balance and motor functions and maintains an appropriate arousal level in the central nervous system as a whole. These morphological divisions, we will assume, are the result of iterated processes leading to the formation of the structures of groups of cells, a process offering rich opportunities for the development of the conflicts that we experience both in balancing our needs against one another and maintaining our position in social life.

But if there is salvation—and according to the religions of wisdom there is—it is to be found in the nature of the process underlying the conflict itself: the iteration of operations in time; in Buddhism, the "turning of the wheel." Just as it is in time that we find the essence of conflict, the alternation of opposing states, so it is outside time that a kind of resolution of conflict can be experienced. An operation that can be expressed iteratively can also be expressed recursively, and just as iteration implies the passage of time, so recursion stands outside time. If it is possible to grasp the nature of that operation and perhaps of all operations as recursive rather than iterative, then they can be perceived as a whole rather than lived through as suffering.

The recursive nature of reality can be clearly seen in someone's life when it is considered after their death. When we remember a dead person, we do not necessarily associate them with any particular age. The dead brother is not the age he was when he died, nor is he the child, nor the young man: he may be none of those things, or all those things at once. We can perhaps see this better when the dead person is also a great person—Rembrandt, say, or Elvis Presley. The sum of such a person's life and output is what characterizes them, not their first, last, or even greatest piece of work. Something seems to emerge from this totality of their life that is the essence of the person. This emergent property of a whole life is allied to the idea of the nature of the deceased and it is not located in time, but is rather the product of all the times that they lived through. However, this defining recursion is also there to be perceived during someone's lifetime, and when we are not engaged by the everyday distractions we can sometimes see it in others or perhaps even in ourselves. This defining character may be seen by painters—it is captured sometimes in portraiture, which embraces many moments together in one view of the subject. This quality of standing outside time is one of the important differences between painting a portrait and taking a photograph. The photograph is localized at one moment, while the painting may also be biographical, and if it is by a true artist, it will be so.

LANGUAGE AND LIFE

So far, we have sought for a mathematical representation of the processes of life. Is there an alternative representation of these processes that also uses iteration as its basic operation and might be capable of giving rise to the geometries of life?

The highest degree of complexity that we know is that of living things and the most complex life that we know of is our own human life. According to the gospel of John, "In the beginning was the Word." This "word" is a form of logos, which has been variously translated as *law, word,* or sometimes *holy spirit.* The use of this analogy in the sacred book of the most powerfully dynamic religion the world has ever seen is highly significant: there is an idea here that must be grasped and understood. The resemblance between the human ability to use symbolic representations such as speech and mathematics, and the idea of the word of God as the fundamental motive power of the world is highly suggestive, and I want to develop this analogy further.

In the course of trying to show the inherent incompleteness of mathematics, Gödel translated the linguistic propositions of logic into purely mathematical symbols. The reverse process is also possible; we can express mathematical statements, however long-windedly, in purely logical-linguistic terms. There is therefore a one-to-one correspondence between logical-linguistic forms and mathematical-symbolic forms. We are thus guaranteed that any mathematical formula or set of formulas corresponds to a set of linguistic propositions. Whatever logical propositions can be expressed in mathematics can also be expressed in language and vice versa.

Can we link these two ideas, one concerned with conflict and contradiction as basic processes and the other concerned with dynamical systems as a creative force? I said earlier in this chapter, when talking about the work of Gödel and Turing, that in both cases what is being computed is a *series* of values rather than one, fixed answer. Plotting such a series graphically generates an attractor, which, as we saw in chapter 5, is characteristic of a dynamical system, with chaotic or strange attractors characterizing chaotic systems.

In chapter 6 we saw how biological shapes closely resemble fractals and in chapter 7 it was shown how shapes of plausible biomorphs can be computed iteratively from simple formulas. At the end of the last chapter we saw also how forms of great complexity can be constructed by the iteration of logical propositions having real truth values. These are typically pairs of propositions that seem to contradict one another. When iterated, they give rise to a chaotically varying set of truth values, the resulting return maps displaying patterns of a strongly organic nature.

"I HATE AND I LOVE"

Our mood is rarely constant for long; it changes in a way that is sometimes regular and sometimes more complex than that. The studies of mood by Totterdell et al. mentioned in chapter 4 show how their changing patterns sometimes follow a simple oscillation (as, for example, with anger) or are chaotic (as with feeling tired).

Is our emotional life translatable into logic? Perhaps—and here is a fanciful example. Let us consider the poem entitled "*Odi et Amo*," the most famous (and also the shortest) by Catullus:

> Odi et amo. Quare id faciam fortasse requiris
> Nescio, sed fieri sentio et excrucior.

In translation:

> I hate and I love. And if you ask me how,
> I do not know: I only feel it, and I'm torn in two.[10]

In the poem, three propositions are expressed:

1. I hate.
2. I love.
3. I feel conflict.

How might we translate this mathematically? Following the spirit of the Totterdell et al. article,[11] we might model the changing mood of the lover by means of the logistic equation.[12] Assigning the variable x to the strength of mood, we have in this case the degree of love represented by the variable x and hatred (its negation) by $1 - x$. The operation "and" is represented by "."

Then given an emotional state, say, $x = 0.7$, that is, a preponderance of love, the next emotional state is given by

$$x' = A.x\,(1 - x)$$

where A is an amplification factor corresponding to arousal. Given a set of initial conditions, we can then calculate the future state of the lover/hater. Let us suppose the lover is highly aroused and $A = 3.8$; a value high enough to drive the iteration into chaos. For the chosen values, the series of emotional states is 0.7, 0.798, 0.613, 0.902, 0.336, 0.848, 0.489, ...

The feeling is first one of strongish love, then even stronger love, then slightly less love (though he is still loving, rather than hating), then extreme love, then suddenly quite a strong feeling of dislike, then love again, and then more or less mixed feelings. This series of alternating emotions will continue and, in fact, the emotional state varies chaotically. Such an approach has been taken by Rinaldi[13] in an analysis of the love relationship of Laura, a married lady, and Petrarch, the fourteenth-century Italian poet and even extended to a treatment of the love triangle of Catherin, Jules, and Jim.[14]

Formulas that produce complex shapes when they are iterated are those that involve unstable values of the variables they contain. We could interpret such formulas in many ways, but if we associate them with propositions, then at least some of those propositions will have variable truth values.

Now let us postulate the following:

Proposition: The forms of life are created by the iteration of processes corresponding to protolinguistic propositions.

Let us consider how much this postulate, if true, would embrace.

It would take account of the essential duality of the world, typified in philosophies throughout the ages, from ancient religions such as Taoism or the teachings of Heraclitus, to modern analyses like those of Descartes, Hegel, and Marx.

It would clarify the emergence of human language: there is a debate in psycholinguistics about whether language ability is restricted to the human species alone. On this iterative view we could say that the capacity for language exists in all life, but that the apparatus for its use varies among species. It is in human language that we see the highest expression so far of this protolinguistic form.

It would help to explain the deep intractability of conflict and the absurdity of hoping for a world in which it is lacking, driven as the system is by fundamentally conflictual iterative forces. All that we can do is ameliorate the effects of this conflict, and to be able to do this we must first recognize and admit its existence. People sometimes say that they hope their lives may some day become happy and settled. This is about as reasonable as expecting that the weather will some day settle down. People are usually more realistic about economics; apart from the occasional extraordinary expression of optimism, people usually expect markets, such as share prices, to go up and down for as long as they exist.

MORPHOLOGY AND CONFLICT

If, as seems likely, the nature of the human condition is intimately linked with its morphology, in other words, if human beings are the shape they are, that shape may best be seen as the outcome of the iteration of propositions with varying truth values. To put it another way, the process that produced our geometry is echoed by the conflicting forces within our natures. Contradiction and conflict are built into the human form and perhaps into all possible living forms. It may not be even theoretically feasible to construct an organism, in the proper sense of the term, without this property of inbuilt contradictions, because it would not be possible to generate it in any other way: an organism lacking iterated contradictions would not become an organized system. In order to create an organism, with the elaborated dendritic systems it contains, it may be necessary to iterate a set of formulas that yield alternating and contradictory truth values; values which, if understood simultaneously rather than successively in time, would indeed be contradictory, but as a result of their continual re-evaluation, produce that quasi-stability of form that characterizes all organisms.

Seen in this light, the forms of biology are the highest products of the world. In no other way would it have been possible for the outcome of iterated logical propositions to be embodied or expressed. Inanimate matter offers the opportunity for simpler formulas to be expressed, those that are the product of addition and subtraction. Others, however, were barred from expression, until the emergence of the complex sequences that form the molecules of life gave rise to the opportunity for multiplication. Only with the evolution of multiplication did something else become possible also; the iteration of the truth values of propositions, some of which are stable and some unstable. Through this the iteration of propositions of a sophisticated protolinguistic form became possible. If the word had literally moved upon the waters of the deep, it could not have found a higher expression of its nature than that made available through the forms of life: and humankind is the crown of that life. This is the message at the heart of iteration.

12

THE WORLD AS ITERATION AND RECURSION

Things taken together are whole and not whole, something which is being brought together and brought apart, which is in tune and out of tune: out of all things can be made a unity, and out of a unity, all things.[1] (FRAGMENT 10)

It is now time to try to gather the threads together and weave them into some kind of whole. Though it may seem to wander at times, the theme is always there. The most basic process in the world is iteration: iteration produces sequence and gives rise to the complexity of pattern that forms the world we perceive. Iteration is the process that governs chaos and generates fractals. Chaos is the dynamic that predominates, with linearity a special case, and fractal pattern emerges from chaotic regimes. The highest expression of these patterns and processes in our world is biological organisms. The animate and inanimate worlds act as real-time computers that, by the manipulation of sequences, produce functions of ever-increasing complexity over time. An inanimate world can produce only additive formulas, and hence only simple Euclidean shapes and amorphous fractals. An animate world leads to formulas involving multiplication, allowing the continued course of the emergent computational properties of the world. The most complex formulas are those computed by biological organisms, the living geometrical embodiment of functions that could not have been produced in any other way. Knowing this, we can better understand thought and perception. Mathematics, the most abstract form of thought, is founded on the same iterative process. The paradoxes that have emerged are the results of applying a timeless mathematics to a time-ruled world; this reaches its most acute contradiction in Gödel's theorem. These paradoxes are only the symptoms of a deeper contradiction, the inherent contradiction of life itself. Since biological forms are the outcome of logically conflicting propositions, our lives, too, take the form of contradiction in thought and conflict in action. This is the reason for the unsatisfactory nature of the human condition, locked as it is into time and iteration.

That, briefly stated, is the message of this book. What I want to do in this final chapter is to examine its implications for a wider view of things, including some areas usually reserved to philosophy and religious thought.

THE UNIFICATION OF THEORY

Throughout this book I have deliberately avoided the use of the word *universe*, because the idea of a universe is greater than, and implies much more than, just the sum of the things within it. To speak of a universe implies that there is an overall organization and unity to the world. We would expect, for instance, that the same laws of organization hold throughout the whole of the world as we have discovered in the part where we have done our experiments. We might expect other things as well, such as the invariance of those laws over time as well as throughout space; a guarantee that what holds today will also hold tomorrow; and we might also expect an answer to the question of how the universe began and how it will end, if it will end.

Scientists have long hoped to develop theories describing things as a whole and have called such theories "universal." Grand Unified Theories (GUTs) and Theories of Everything (TOEs)—these grand names, with ridiculous-sounding acronyms—describe a simple idea. They refer to theories that say how the fundamental laws of nature can be unified. Such theories, it is hoped, could be written in a simple form, perhaps as a single equation that predicts the next state of the world, given the present one.

It might seem that a universal theory, far from being an explanation of everything, could only explain broad outlines. However, such a theory, if it could be found, would imply something far wider. If we knew the basic laws of the world (plus the conditions from which it started), we could describe all of its history and forecast the future course of its evolution in some detail. Another tendency is also at work in the search for these universal laws: the desire to be free of detail and to be able to unify phenomena at some higher level. There is a widespread feeling that physical phenomena and physical laws have multiplied in complexity beyond our present ability to put them together into one comprehensible picture: that it is time for a great simplification.

SPLITTERS AND UNIFIERS

The search for a theory of everything is the latest manifestation of a phenomenon that has recurred in epochs throughout the history of science. Isaiah

Berlin divided scientists into two camps—splitters and unifiers—and this is reflected in those historical changes. At times there has been a tendency in science for knowledge to split into many fragments and disciplines, each dominated by its own ways of thought and its own academic and professional bodies; at other times there is a unifying tendency, when many different strands of knowledge are brought together, perhaps by some revolutionary new theory or by some other factor of social change or spiritual insight.

Four hundred years ago Francis Bacon proposed that science advances by accumulating facts out of which a new hypothesis somehow emerges. This is called the inductive method, in contrast with the deductive method of deriving predictions from theory. Since Bacon's time many people have thought that principles of induction could be found that would enable new hypotheses in science to be formulated (and scientific understanding to advance) along orderly, well-planned lines, but no one has ever succeeded in formulating these principles, if indeed they exist. Other factors have been pointed out by the so-called sociologists of science as much more influential in the emergence of hypotheses: these include the social background in which research is done and even the individual mental outlook of the scientist. In any event, science does not seem to progress in the orderly way Bacon suggested.

The swings between periods of detailed discovery and of theoretical unification fit much better into the view of Thomas Kuhn,[2] who suggested that science follows the same broad method (paradigm) until it fails and a new paradigm emerges. A theory of everything would end this series of alternating approaches to knowledge because, if it were found, everything would then be known.

Yet, if the theory of everything were ever found, it would still not be enough to define the totality of things as a universe The fact that we can predict the course of events does not make them into a system, any more than the ability to enumerate them does. In chapter 2 we saw that a system such as a biological organism has properties when it is considered as a whole that cannot be predicted in any way from even a complete knowledge of the interactions of its parts. Even if everything in the world can one day be described and predicted, we can never know anything about it as a universe. That would be like trying to predict what an organism will do when all we know is its physiology.

FLATLAND

The term *universe* is deceptive, since it implies that we know the properties of a typical universe. But we could only know these by getting outside our universe to find out what those properties were, and this by definition we cannot do. We can't, for example, study the behavior of a *population* of universes, so we know nothing about how a particular universe should behave. It looks as though we can never know that the universe exists, and we would be forced to be content with the study of the purely internal events of the world.

If we are to speak of a universe, we must find some way of redefining the totality of things to be a whole, but it is not obvious how this can ever be done. Part of the reason is implied by the view of iteration as a fundamental process. Iteration produces small changes and these accumulate to produce large changes over time. We feel we can observe this process, but we are also a part of it. Because we can observe it, we can see things changing around us all the time, but we do not see ourselves as objects that are changing, because we can only see ourselves from within. For the same reason, we cannot know anything about the world as a whole. That would require us to be able to observe the whole world from the outside and it is impossible for us to make such observations.

George Spencer-Brown has illustrated this kind of problem very clearly in his little-known but seminal book *Laws of Form.*[3] (In giving this account I am deeply indebted to Robin Robertson for his commentary on Spencer-Brown.[4]) In this book, Spencer-Brown imagines a creature crawling about on a plane surface like a tabletop. This creature is endowed with enough wits to be able to notice when it crosses a boundary, but to begin with, it knows nothing more.

The surface is at first undivided and can be thought of as a plenum of possibilities, from which so far none has emerged. To create anything from this plenum we must make what Spencer-Brown calls a distinction. The most basic way of doing this is to draw a closed curve (equivalent topologically to a circle) on the plane as in figure 12.1.

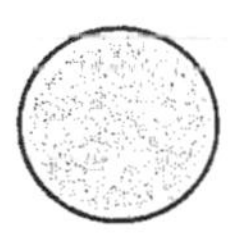

FIGURE 12.1 A distinction in the plane.

This creates two areas, inside and outside, plus and minus, or, in general, any two opposed states. Seen from the creature's viewpoint, the world consists only of these two states created by the distinction, and an event—a transition—when it passes from one state to another.

Suppose it enters the circle drawn on the plane (figure 12.2). Then it knows (by having made a crossing) that it is inside rather than outside.

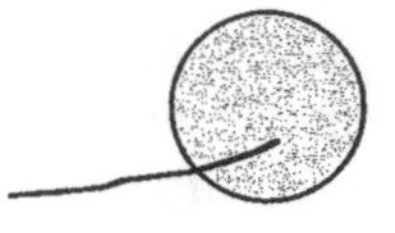

FIGURE 12.2 A crossing.

Now suppose there is another circle (distinction) in the plane. The creature can never know this. If it enters and leaves the first circle and then enters the second circle, then as far as it knows, it might have left and then entered the first circle for a second time. So two circles are, in the creature's system, equivalent to one circle, as in figure 12.3.

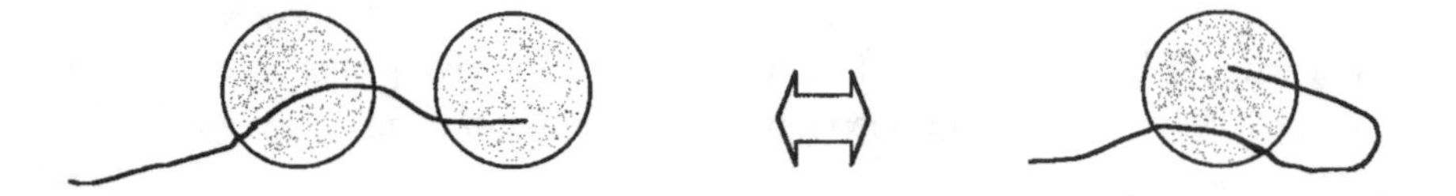

FIGURE 12.3 The equivalence of two adjacent distinctions to one.

The next case to consider is that there are two circles (distinctions), one inside the other. Suppose the creature enters the outer one and then immediately enters the inner one, as in figure 12.4.

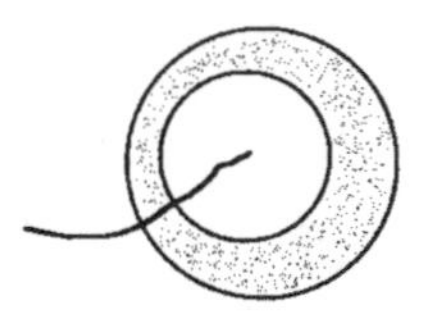

FIGURE 12.4 Nested distinctions.

It knows it has crossed two boundaries, and when it finds itself outside, it thinks it has entered the first circle and then left it again. So two concentric circles are, to the creature, equivalent to none, as represented in figure 12.5.

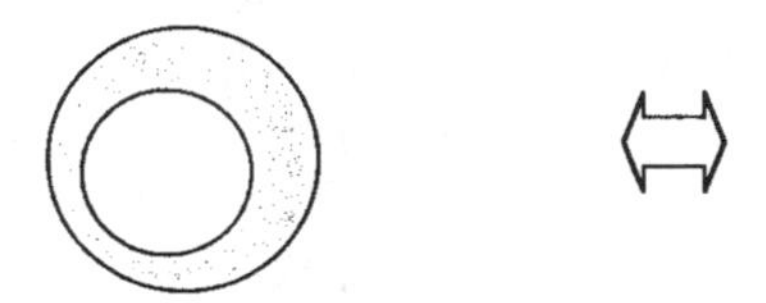

FIGURE 12.5 The equivalence of two nested distinctions to none.

In this simple system any sequence of boundary crossings can be reduced to either one or no crossings. Spencer-Brown goes on to make from this model world a series of symbolic statements so powerful that their significance has still not been widely appreciated.

Now, the point to note is that the creature cannot count. In its system all that it remembers is one event—the last one—and from this it knows whether it is inside or outside. Its mind is, in effect, like a switch, without any memory of its past states. In other words, it lives iteratively. The creature's knowledge of its world is incomplete: the result of its perception is not an accurate portrayal of its world.

In order to have a memory it must enter a more advanced world than the plane. If the creature is to remember the number of times it makes a crossing, it must be able to leave its plane and move through a third dimension,[5] arriving back on the plane through a sort of worm-hole like that shown in figure 12.6. This allows it to see, rather than simply feel, whether it is inside or outside.

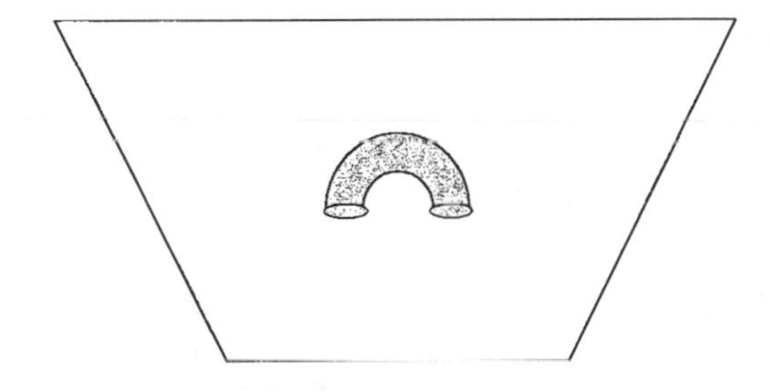

FIGURE 12.6 Leaving the plane.

What has happened to the creature in acquiring a memory? It has also acquired the ability to count. But it cannot both count and remember at the same time. As Spencer-Brown puts it, "counting is the contrary of remembering." The creature is condemned to live either iteratively or recursively.

Yukio-Pegio Gunji, an earth scientist working in Kobé in Japan, has discussed the question of outside versus inside in the context of self-measurement in systems. Gunji looks at a very basic level of self-description in a system. In a number of papers[6] he talks about the distinction between internal and external measurement. A system can make internal measurements of itself and in that way can define its own time. However, according to Gunji's view, a complex system cannot make an external measurement of itself. Thus each system is in a sense isolated. So when a system tries to measure itself externally, it is involved in what he calls a recursive (i. e., iterative[7]) problem. The solution to this problem is the existence of a fixed point that implies a contradiction[8] and is thus a paradox. Such paradoxes include Turing's version of the *entscheidungsproblem,*[9] Gödel's theorem of undecidability, and fixed-point theorems such as Birkhoff's for lattices.[10]

Gunji and Spencer-Brown are expressing the same point: that a system cannot adequately describe itself. In order to fully describe a system we must be able to get outside it. As Wittgenstein said on the subject of boundaries, we can see *to* the boundary, but we cannot see *beyond* it, nor can we see the boundary from outside. Applying this to time rather than to space, he said "Death is not an event of life. Death is not lived through. . . . Our life is endless in the way that our visual field is without limit."[11] We cannot talk about the view of the outside of our world any more than we can describe ourselves as being dead. In order to describe something as a system we must be able to stand outside it and observe it. Clearly, with the world as a whole this is impossible.

FINDING A WAY OUT OF THE WORLD

Or is it? If it were possible to stand outside the world, we would have to be able to look at it from outside space and time. Is there any way in which this might be achieved?

Mystics would say that there was. The mystical approach is a contemplative one, and mystics generally take a holistic view of the world, asserting its essential oneness. Could there be something about the contemplative approach that yields knowledge not available to observers using the senses alone? Moreover, could we somehow leap from the particular to the universal and say something about the world as a whole?

There has been a tradition in Anglo-Saxon philosophy (which is largely empiricist philosophy), and the science that follows from it, to reject any path to knowledge other than through sensory experience. It is a cardinal point of this tradition, which includes Locke, Hume, and, in the twentieth century, Wittgenstein in his early period, that all knowledge comes from the experience of external reality. The main argument used against accepting the findings of mysticism is that sensory experience is the only kind that reflects the state of affairs in the outside world. If we accept this, then someone who is having what mystics call a direct experience of reality is not perceiving anything in the usual sense of the term; that is, they are not receiving information through the senses. In addition, when mystics are aware of their own mental experiences, these are likely to be influenced by factors entirely separate from external reality, perhaps by habitual or received ideas or by a desire for some kind of mental satisfaction. During the states that traditionally give rise to mystical experiences (meditation, isolation, drugs), the brain is largely cut off from sensory input and in such states it may behave more like it does during hallucinations, which are notoriously misleading. So, it has been said, we should reject mystical insights, including those revealing the unity of the world, and limit our statements to what can be learned from systematic observation.[12]

However, such mystical insights are reflected in physiological activity. A brain area (part of the posterior superior parietal lobe known as the orientation association area or OAA) that becomes much *less* active during religious contemplation of the kind usually thought of as mystical has recently been identified by Newberg, D'Aquili, and Rause.[13] It might seem reasonable to associate this area with a religious perceptual sense, in the same way that the primary visual cortex is associated with the perception of visual objects. Is not the OAA perceiving something that is just as real as in the case of visual perception? Newberg, D'Aquili, and Rause suggest that this is indeed so and that their research indicates that "the mind's machinery of transcendence may in fact be a window through which we can glimpse the ultimate realness of something that is truly divine." Critics in return dismiss these speculations on the grounds that "Delusion is simply a description of what happens when the OAA shuts down and the brain loses the ability to distinguish self from non-self."[14] What are we to make of this debate?

There are two answers to the classical objection to knowledge by direct awareness, the first epistemological and the second ontological. Everything we know about the external world is known only as a mental event. Since our entire knowledge of the external world is based on mental experience, to reject that experience is to reject the whole of what is known. In fact, mental phenomena are the stuff of a psychology whose purpose is to explain perception, cognition, emotion, and other aspects of mental experience. Again, a mental event tells us something about the person experiencing it and perhaps something about mental events in general. It would pay us to be cautious about rejecting as inaccurate, unacceptable, or unreal any psychological data, however subjective they may be.

PERCEPTION AND IMAGINATION

The distinction between internally and externally gathered information is sometimes hard to make. If someone is asked to imagine an object, say a banana, and simultaneously shown a real banana, they may combine the two perceptually, so that they are hard to disentangle afterward. Cheves West Perky[15] in a series of experiments in 1910 asked people to imagine simple objects while looking at a blank screen. The experiment was done in the evening when the laboratory was quiet and the participants thought that only they and the experimenter were there. What the observers did not know was that an assistant in the next room was projecting faint pictures on the screen that coincided, and sometimes conflicted, with the objects being imagined. The observers' reports showed that they were utterly confused as to what they had imagined and what they were being (unknowingly) shown. When the slide was the same as the mental image, observers were convinced they had imagined the whole percept. Comments such as "It is more distinct than I usually do;" "Feels as if I was making them up in my mind;" or "I get to thinking of it, and it turns up!" clearly demonstrated their confusion. When the picture and image were different, then a combined image often resulted. A subject asked to imagine an elephant was shown a slide of a cup of tea and this yielded a merged mental image of a cup of tea with a worm in it! It is well-known to perceptual psychologists that if two different faces are shown, one to each eye in a stereoscope, the observer sees a combined face. We are usually unable to tell whether a given piece of information comes from ourselves or from outside, particularly after it has been combined with other pieces into one perceptual experience.

Experiments have also shown how a perceptual act consists of comparing the internal and external information.[16] Internal images are used like prototypes against which we compare the incoming signal. If signal and image match, then we perceive correctly; otherwise we are either hallucinating or incorporating the outside world into our image, as dreamers may incorporate sounds into their dreams.[17]

Where do these prototypes come from? According to the empiricist view, they come from past experience. But that answer will not account for the innate abilities of most species, including humans, to perceive certain objects from birth,[18] or for their sensitivity to stimuli, such as sexual cues, at particular periods and without prior experience. The evidence is overwhelming that many mental structures are innate, including those on which we base our perceptual world. Although we can generate new cognitive structures of certain types, we seem to be born with the ability to represent knowledge only in particular ways and we apparently cannot think outside those limits.[19] All this suggests that the brain is replete with structure from birth; so at least some of this structure and the information it stores can be perceived from within, by direct experience.

THE MAN WITHIN

What, then, happens during mystical experiences like meditation? The common techniques of meditation involve continuous contemplation of an object: sometimes this is the repetition of a mantra, or the visual repetition of looking at an object or of contemplating an internal image, or becoming aware of breathing, or simple awareness itself. There is little similarity between these states of meditation and those conditions of pathological hallucination that often involve organic damage or toxic factors. There is no suggestion that mystics are similar to schizophrenics, paranoid personalities, or other mentally ill people. There is much more similarity to the experiences caused by some drugs, particularly psychedelic or mind-expanding drugs, such as mescaline and LSD. All drugs are clearly not in the same category. Some, like alcohol or benzodiazepines, depress the nervous system and blunt perception, whether internal or external, while others, like cocaine or tea, stimulate it. Psychedelics are even more involved in their mode of action. It is claimed by those taking psychedelics that through them they experience an expansion, not a contraction, of awareness. However, the class of psychedelics is itself diverse, and the effects of LSD are essentially different from those of other drugs labeled psychedelic, such as mescaline or psilocybin. Some of these effects are peripheral. For example, LSD increases the intraoptic pressure, which may be at the root of some of its reported hallucinatory effects. Other effects are more central and act via the chemistry of the brain.

These effects can occur only because there are receptors for these drugs already present in our bodies. There are receptors in the human central nervous system for cannabinoids, opiates, nicotine, DMT, adrenochrome, and many other psychoactive substances. Many of these are also produced endogenously under appropriate circumstances, for example, in response to fear, injury, exercise, relaxation, and meditation. The human endocrine system is a self-contained and perfectly legal drug factory. We internally manufacture our own morphine-like substances, hallucinogens, psychedelics, and stimulants and we respond to them in a targeted and sensitive way. There is every reason to think that what can be done moderately badly by an externally administered drug can be done much better by the substances to which evolution has adapted us, if we can only gain access to our internal pharmacy.

During the act of concentrative meditation, the mind becomes focused on a sensory input such as breathing, or an internally manufactured image like the repetition of a sound. In the technique known as relaxed awareness, the mind is defocused and is aware simply of being aware, the aim being to empty the mind as far as possible. In its highest form this technique extends to the contemplation of the void, in which the world is, in imagination, emptied of objects, ending finally with the meditator themselves. If the object of contemplation is a repeated sensation, then as we adapt to that sensation we become aware of the act of perceiving rather than the thing perceived. In this way we gradually become aware of our own mental processes. A further element is that, because the repetition of the stimulus is a form of iteration, we become aware of the iterative nature of the mental acts of thought and perception. The end result is that, whatever kind of meditation is adopted, a heightened sense of awareness is almost invariably achieved. Is it possible that this mental state is able to convey some knowledge about the nature of awareness itself?

The nature of consciousness is still a mystery, but has never been of more fascination to scientists, whether psychologists, physiologists, or physicists. While there may be disagreement over the epistemological status of awareness, people are generally agreed that it is a state involving self-referential activity. Consciousness involves the knowledge that one is conscious; it is a reflective state that even reflects itself, and we could picture it as some sort of mirroring process. This is one aspect of the age-old puzzle of what consciousness is. An old-fashioned way of picturing awareness was to propose that our sensations are perceived by a homunculus (little man) located inside our head who views the sensory input coming in to the brain like someone watching a television set. The question naturally arises of how the homunculus does his perceiving. Does he have eyes inside his head looking at his incoming sensations on a yet smaller screen? The problem was in this way pushed a stage further back: you must then propose that there is a second homunculus inside the head of the first one, and so on in a seemingly never-ending sequence, like Russian dolls. This was always considered a conclusive argument against the homunculus idea, but seen in the light of a post-Gödelian mathematics, such ideas lead instead to a recursive view of consciousness. Consciousness is then in some sense a reflection of itself; like mirrors facing one another, an infinitely continued iteration.

ENTERING THE RECURSION

As we have seen, recursion and iteration are inverse expressions of the same process. Faced with trying to express an iteration, we would usually prefer, for brevity, to express it as a recursion. Then, instead of trying to write out all the iterative steps, and in the end leaving a string of dots to show what we cannot complete, we can write a simple, self-contained formula. The reason for the two different forms of expression is partly (though not simply) notational: in order to include time in the mathematical notation, we need to use the recursive relation; if time is implicit, as in computing, then the iteration is more natural.

Herein lies the key to a part of what consciousness is. If we are to grasp the nature of consciousness, we must somehow step out of its recursive nature and perceive it as an iteration. That can be difficult, and involves another reversal, for this time the inverse confusion is occurring. The process of being aware appears normally to be a recursion because we stand outside of time—the physical time that causes the progression of steps in the iterative world. By becoming aware in meditation of the iterative nature of our consciousness, we come to perceive it from outside, as a system. We move from the automatic recursion that is our natural everyday life to a perception of iteration in the act of self-awareness.

THE STUFF OF REALITY

The question of the relation between consciousness and the world is an old one. According to idealists, consciousness is the basic stuff out of which the world is formed. The materialist view is the inverse of this; it sees the dependence of mind on matter as proving that matter is the fundamental substance. These two philosophies are mirror images, since both idealists and materialists are agreed on one thing—that there is only one substance. Just as the statements in the paradoxes contradict and at the same time affirm one another, so the idealist and realist views complement one another through their apparent mutual contradiction.

By seeing the world from outside as an iterative system, meditation can indeed give us knowledge by direct acquaintance with the world as a whole, and this is why mystics have cultivated the sense of unity that is the ultimate reason for the idea of a universe. All of us have this sense, more or less highly developed. Scientists, too, frequently have a sense of the mystical (though they often would not like to put in this way). They, too, are aware of the wholeness of things, which may be one reason why they have so readily adopted the idea of a universe.

In the same way, the everyday perception of reality gives us a seemingly eternal view of things (There is only now!). For these reasons, there can never be conclusive evidence for the existence of a universe from the observation of sense data by themselves. Such observational evidence may be suggestive: the clinching evidence can come only from insight gained through contemplation.

THE CONSTITUENTS OF THE WORLD

The conventional approach to the question "What is the world made of?" is the atomic one. We cut up matter into smaller and smaller pieces until, we hope, we can arrive at the atom, which, like a child's toy bricks, is the ultimate building block from which everything is made.

The search for the atom has gone on since the beginning of science. It was Democritus[20] who was first recorded as saying that nature could be split into component parts of a limited number of types, with each of the same type identical. In the mind of the scientist there often seems to be a memory of a construction kit played with as a child, or the building blocks piled up in endlessly different ways. Of course, building blocks are not identical. They are made from different pieces of wood and have slightly different shapes. What matters is that they can be substituted for one another; one block is equivalent to any other. These are the atoms of the play world.

When we move from the world of play into the world of science, we may continue to look for the atoms of whatever science we are dealing with. The word *atom* means "indivisible one"; an atom is the simplest thing that can be found. If we have really found it, it should be possible to build any object out of a sufficiently large number of atoms. But this doesn't happen, for three reasons. First, we can't be sure the thing we have labeled as an atom is in fact indivisible; it frequently turns out to have parts even smaller than itself. Second, the atoms are far from simple. Indeed, the smaller they get, the more complex they seem to become, and the less predictable their behavior. Third, it is not possible to use atoms to predict what will happen when they are combined: emergent properties are not adequately explained by atomic theories.

Two familiar examples will serve to illustrate this last point. If we combine sodium atoms with chlorine atoms under the right conditions, we will get common salt. Salt is edible, harmless in reasonable amounts, tasty, and necessary for life. Chlorine and sodium are none of these things. One is a gas that will kill if inhaled; the other is a volatile, corrosive solid that must be kept well away from water. It is not possible to predict the properties of salt from those of its constituent elements.

My second example is from human psychology. If we invite a number of people to a dinner party, we can't predict what will happen. We may know them all very well separately, and get on well with them on this basis, but bring them together, and who knows what will happen? Dinner parties are going out of favor in many places, because of the stress they can engender—a property of the mixture unforeseeable from the ingredients.

It is not even always possible to predict the outcome on a new occasion, having once done something. We don't get salt, or we get something else as well; the experiment "didn't work." We invite the same people who got on so well together last time, and they have a row with one another.

There is a school of thought that says this is a not a problem, that we can predict properties of the whole from those of its constituent parts. In the case of "new" molecules, for example, their properties are predictable from their chemical bonds and molecular weights. A theory of emergence was thought to be necessary in the early part of the twentieth century to explain chemical properties, but the theory of chemical bonding superseded it, since it is claimed that the structural properties of new chemical combinations can be predicted by the strengths of their chemical bonds.

Two objections undermine this approach. First, it does not always work, in the sense that the properties of the parts do not always bear a known relation to the nature of the whole. Chemical affinities, for example, will only partly explain the mechanical properties of new substances. But a more important objection is that, even if these properties could be always explained, this really evades the problem, since chemical bonds did not exist at the time of the Big Bang, but had to await the condensation of matter to come into being. In an evolving universe new properties will emerge as a result of the interaction of simple elements, and the first time this happens, something has emerged. To say that we now can predict the conditions under which higher level properties will appear is to speak post hoc. When the property emerged for the first time it was not predictable, if only because there was no one there to predict it.

FROM ATOMS TO MONADS

What is the resolution of this paradox? If something new emerges from the combination of familiar elements, where has it come from? It cannot come from nowhere; the only remaining possibility is that it must have come from the units themselves. The dinner party guests must *contain* the possibility of a pleasant or an unpleasant evening; the chlorine atoms and sodium atoms must *contain* the properties of salt. The seventeenth-century mathematician and philosopher Gottfreid Leibniz, in his essay "Monadology,"[21] suggested that the fundamental unit of all things is the *monad*. He intended the monad to have some of the attributes of the atom, but with important differences. The monads Leibniz proposed are indivisible, indissoluble, and have no extension or shape, yet from them all things were made. He called them "the true atoms of nature." At the same time, each monad mirrored the universe.

If each monad contains all the properties in the universe, it is itself a universe in miniature, so when we manipulate monads in certain ways, any property in the universe may appear. Suppose we have a number of purses containing a variety of items, for example, money in the form of coins, notes, or precious articles. The purses have zippers, so that when we move them around nothing falls out. If we unzip some of the purses, some properties can emerge. We may be surprised, if we think the purses are empty, first, that anything appears, and second, that different things appear when we carry out different operations on the purses (but that would be because we knew nothing about the nature of purses). Which particular things emerge depends on the way we treat the purses. If we shake them up and down, coins will jump out and we will get a collection of different coins; even if all the purses contain the same coins, we will get a different combination of coins every time unless we shake them in almost identical ways. We might then form a theory that purses were made up of coins. By contrast, if we blow into the purses, notes will fly out. We might then declare a second hypothesis: that purses are made up of notes. We might in puzzlement even propose a duality hypothesis, that purses are sometimes made of coins and sometimes of notes.[22]

This is an imperfect analogy perhaps, but in the same way, rather than mysteriously emerging as a result of combining larger and larger units, the properties we observe might be better conceived of as contained within the monad. What then corresponds to the different treatments of the purses? I have proposed[23] that it is the order or arrangement of the monads that causes particular properties to emerge. Leibniz intended that no two monads be alike, on the grounds that no two beings in the world are precisely alike, but I suggest that all monads be alike. If identical monads are arranged in different ways, then diversity can appear. And if all monads contain all properties, then the property that emerges effectively becomes a function of the arrangement of the monads. We already have a model of objects that contain their own properties—fractals. As we saw in chapter 4, a fractal is self-similar and it has *nothing but* arrangement. Any part of a fractal contains the properties of the whole; this is indeed the definition of what a fractal is. It contains itself, and its detail is infinitely small ("having no extension").

TRUE ATOMS

It may be objected that what we have here is not an atom at all, since there is no ultimate constituent, but only an infinite regression giving rise to many levels of a phenomenon, and moreover, one that starts at the wrong end—with the macroscopic—and works down toward the invisible. But the infinite regression is in fact a good model of the levels of objects with which science deals: atoms, molecules, cells, persons, and so on. By iteration alone we are not apparently defining any single constituent part. However, we can bypass the iterative levels by restating the iterative relation recursively. The monad can then be defined as a set of recursive properties—in fact, the set of all such properties. Each property then becomes a recursion and emerges as a result of a particular sequence of arrangement of the monads. However, we may note that there are going to be no new properties, since all are already contained within the monad. So the rule for combining properties must not allow for any *new* properties to be produced. One quite flexible way to do this is to specify that the monad is a *superposition* of properties. Which property emerges depends on an arrangement of monads, operating on the universal monad itself.

Because (in this model) all monads are identical, the monad is not a variable but a constant; only sequence determines the emergence of properties. In fact, the property is not new but is there all along, concealed within the monad.

THE IMPORTANCE OF SEQUENCE

I will give two practical examples of the importance of sequence, one from chemistry and one from engineering. The first, mentioned in chapter 1, is in catalysis. Homogeneous catalysis by metal complexes plays an important part in chemical and biological processing, including the production of millions of tons of chemicals and foodstuffs, such as polyunsaturated fats, every year. The Langmuir–Blodgett process consists of passing the reactants to be catalyzed over thin films of a rhodium complex. In a study of the efficiency of Langmuir–Blodgett films, Töllner[24] and co-workers found that the order of the rhodium complex molecules (their orientation in the substrate of the catalyst) was critical to the efficiency of the reaction. They also found that the efficiency increased with temperature, presumably because a higher temperature allowed more rearrangement of the molecules to occur.

The second example is from aviation. One way to improve the efficiency of an aircraft is to decrease the amount of drag on its wings and fuselage. It was found by Sirovich and Karlsson[25] that if nodules are placed on the outer surfaces of an aircraft, they will disrupt the turbulent flow of air around the skin of the craft—the factor that produces drag. The interesting finding is that the nodules are less efficient if they are distributed in regular rows; the drag is reduced more if they are arranged randomly. Just what does randomly mean? It means a sequence whose properties the authors have not shared with us, presumably because they are not known, even to themselves.

Findings like these are indications of the importance of order and sequence effects in some of the most basic physical and chemical processes, effects that are just beginning to be appreciated. Such questions can only be tackled by a consideration of sequence and its meaning.

EMERGENCE AND EVOLUTION

The interpretation I have suggested implies that the only properties that can emerge are those that can be produced by a given sequence. This may help to explain why more and more properties appear as the universe evolves. At first, when there were only short sequences possible, only a limited set of properties could appear. As the universe grows, more and more properties can emerge, because sequence lengths can grow also. For example, small particles were the first to appear, the free movement of electromagnetic radiation came later, condensed matter formed later still, and so on, culminating in the long molecules basic to life, in which sequence is all-important.

If there are a theoretically infinite number of steps involved in each recursion, why do we observe only a few layers of emergence: the physical, the biological, and so on? We can only observe down to a certain level corresponding to the energy we use. It requires more and more energy to get to a deeper level. Simple substances can be split into their component chemicals by applying heat, and this was the fire revolution. If you apply more heat, you get more rearrangements, until you arrive at what were prematurely called atoms. To split the atom required a temperature of about ten million degrees Celsius. To split subatomic particles in an accelerator requires energies of the order of hundreds of billions of electron volts (GeV). If we could go deeper than is presently possible, we would discover (as we probably shall) further layers that are at now hidden because of the energy needed to reveal them. But we would never find a fundamental particle, because only monads, not particles, can be fundamental.

Emergent properties then become explicable, at least in principle. If each constituent of the world (the monad) contains all the properties of the world, then any of these properties may emerge as a result of the combination of these constituents. The particular properties that emerge will result from the conditions of combination. There is nothing new in such a world: emergence is simply the revealing of what is always present. An apparently new property is created by the combination of elements already containing that property. Thus the property arises as a result of the property. This implies that a given property is defined ultimately in terms of itself, just as a fractal form is defined in terms of itself, in a recursive manner.

THE ONTOLOGICAL ARGUMENT

Does what applies to properties in particular also apply to properties in general, or even also to the totality of properties? The ontological argument for the existence of God was traditionally one of the Catholic Church's three "arguments from reason" (i. e., those that do not require revelation) by which God's existence can be proved. Can the existence of something be defined in terms of itself?

The ontological argument is that God must exist because He is the perfection of all properties and therefore possesses the property of existence also. Bertrand Russell, a notorious skeptic, but one of the foremost logicians of his time, said that when he suddenly realized that this argument was sound, he nearly fell off his bicycle. Could we say it was sound according to this monadic view?

Whether existence is or is not included as a property depends on what we consider a property. If properties include only things like "green," "square," and "round," then "existing" seems very different. If we include "electrical," "abstract," and "recursive," then we come closer; these are certainly properties, though not immediately tangible ones. But the reason why "existence" is often excluded from being a property of something is that it seems to add nothing to what is already being said about that thing. If we allowed the property of existence, then by using a variation of the ontological argument, might we not prove the existence of anything we liked, including such bizarre objects (or nonobjects) as flying saucers, unicorns, and ghosts?[26]

When it comes to the totality of things, as with the universe and with God, it seems impossible to say anything about their reality from the evidence we have, that is, the sensory evidence that is the starting point of science. Only on the basis of direct experience could we say that there are or are not grounds for belief in them. People's experience, especially their religious experience, varies: sometimes they believe in God, sometimes they have doubts of the deepest kind, and these doubts are not limited to believers. There can be very few religious people who have managed to sustain a lifelong belief in what they have been taught, or few atheists who have not sometimes doubted their own skepticism, if for no other reason than that it is psychologically impossible to keep a fixed belief forever. Perhaps these people underestimate God, thinking Him a fixed object. If we accept the iterative–recursive apposition, then perhaps He exists and does not exist, or perhaps He does not exist all the time.

SYNCHRONICITY

In this book I have argued that because science is the source of our logos, since science is at an impasse, we also are at an impasse. There is no new direction for analytical thought to take. Within the present scientific paradigm, whole categories of events are rejected from consideration because they do not fit in with that paradigm.

One example is the occurrence of coincidences. It is everyone's experience that remarkable coincidences often do happen. Everyday examples are meeting or hearing from someone you have just been thinking about or finding the same word or name cropping up in several different places in a short space of time. Many people feel that at times their lives are permeated with these striking and apparently significant events. However, if scientists are asked for an explanation of why these events happen, they dismiss them as "mere" coincidences. In a random-selective world, a coincidence remains mere, because no causal relationship between the events comprising it can be found. In an iterative-sequential world, on the other hand, coincidences will happen by virtue of the self-similar structure of such a world. The same things will be expected to happen at the same time in different places; if they did not, the world would not be iterative.

Monozygotic (identical) twins share a common genetic inheritance. They are occasionally separated at or shortly after birth and they often share a common fate in their separate geographical locations. These coincidences are often reported in the newspapers and might be dismissed as popular myths, but scientific studies also exist. Dr. Bouchard of the University of Minnesota surveyed the lives of separated monozygotic twins and found cases like that of James Springer and James Lewis. Each adopting couple believed that the other baby had died. In addition to both having the name James, both twins married a woman named Linda, were divorced, and then remarried a woman named Betty. As children, both liked mathematics and hated spelling and both owned dogs that they named Troy. Both worked as filling-station attendants and for MacDonalds hamburger chain and were at one time deputy sheriffs. Both suffered tension headaches that began at 18 years of age and proved to be migraine. These attacks started and stopped at the same age in both brothers. As well as sharing many physical and behavioral characteristics, they scored almost identically on psychometric tests.

EXPLAINING, OR EXPLAINING AWAY?

Such cases are remarkable, but may seem explicable because there is a mechanism—in this case a shared genetic structure and possibly shared early memories—that might account for these common features. However, sometimes the coincidences seems to be beyond the power of known mechanisms to explain them. A case reported in the *Newcastle [England] Chronicle* concerned two brothers who were identical twins, one of whom had emigrated to Australia while the other remained in England. In addition to the usual shared details of life history, both twins died within a few hours of one another. After the death of the brother in Australia, the survivor in England, who was already ill, asked "Why is Frank standing at the bottom of my bed?" just before he died.

In a world modeled on random-selection principles such phenomena are not only inexplicable, they are annoying, because they stand outside the established paradigm and strongly suggest its inadequacy. Explanations of coincidental phenomena are often attempted, usually along the lines of trying to apply probability theory. How many instances are there of coincidence occurring by chance? This ought to be a function of the total number of possibilities. If two sets of circumstances have many hundreds of characteristics, it is argued, then they will by chance alone share some of these.

This kind of explanation can be used to question the validity of any claimed extraordinary event, for instance, precognitive dreams. People often dream of what will happen the next day. Sometimes the events do happen and in a way that excludes the possibility that the prophecy could have been self-fulfilling. In the most dramatic cases these dreams are of disaster, obviously outside the control of the dreamer. The argument against their significance goes like this: Many dreams occur in a single night over the entire area from which they might be reported. Some of these will surely be of disasters. If a disaster later turns out to have happened, is this not a mere coincidence, without any systematic link between the dream and the event, and due to the operation of chance? If a calculation can be done that yields an appropriately low probability of such an event, then in this way the random-selection paradigm can be preserved and any phenomenon that does not conform to it can be explained as coincidence. But probabilities are only of use in the absence of more precise knowledge. So to take the approach of calculating probabilities is to already assume there is no connection, as in the random-selection paradigm there cannot be.

Now let us look at the phenomenon of coincidence from a different worldview—that of the iterative-sequential paradigm. In such a world, self-similarity is a basic property of things: in a self-similar world the same features develop at the same time in different places. Coincidences will therefore be expected events. This is the reverse of the case of the random-selection paradigm. In such a world an explanation is required if there is *not* a coincidence, because the absence of coincident patterns implies the existence of a systematic disruptive external influence.

To take an example from a different area, quantum mechanics shows that coincidence operates at a very fundamental level. Two light particles emitted from a common source can become entangled so that they share reciprocal properties. What happens to one particle will affect the same measured property of the other. This happens even though the particles are moving apart from one another faster than the speed of light and the effect can be used to transmit information from one place to another at supraluminal speeds.[27] In an analogy with this quantum entanglement, Grinberg-Zylberbaum[28] tested pairs of people who had established rapport by meditating together. When one was shown a randomly flashing light, the brain of the other could exhibit transferred potentials like the evoked potentials in the brain of the observer.

THE PREESTABLISHED (DIS)HARMONY

Preechoes of the iterative-sequential paradigm can be found in views of the nature of the world that came long before quantum mechanics. One was the *preestablished harmony* of Leibniz, which was an important part of his monadology. The idea of preestablished harmony was an adaptation of the earlier ideas of Malebranche and Geulincx, who, in the seventeenth century, had independently suggested the notion of a synchronization between the brain and the mind.

If, as they appear to be, the mind and brain are two different substances, the apparently almost exact correspondence between the two is a puzzle.

Malebranche and Geulincx's answer to this problem was: if two clocks are set running at the same time, and if both are accurate, then they will continue to tell the same time even though there is no connection between them. This, they suggested, was like the relation between the body and the soul (or as we would put it, between the brain and the mind) and accounted for their otherwise inexplicable correspondence. They had been set running together, at the time of conception, by God.

Leibniz adapted the argument for a different philosophical purpose, as part of his explanation of the world of conscious beings. In his theory, the world consists of centers of consciousness (monads), which are essentially self-contained and isolated from one another, just as our minds seem to be. In order to account for their ability to communicate despite their isolation, Leibniz assumed the monads to be in a relationship of preestablished harmony, like a multiplicity of clocks in the Geulincx–Malebranche model. In this worldview, our separate consciousnesses are in a preestablished harmony like so many clocks, so that what you think coincides with what I think, although we are separated, and this amounts to communication. In this model it would be no wonder that things happen to us coincidentally, just as the clocks in a clockmaker's shop all chime twelve together, or all show ten minutes past two together. It is another way of saying that if monads are the set of all properties, then each contains the properties of all the others.

The philosopher Alfred North Whitehead (the other half of the Russell–Whitehead *Principia Mathematica* partnership) took the view that reality consists, not of atoms, but of organic wholes. His was not a fashionable way of looking at things and the atomism of his partner and erstwhile student prevailed in most minds. But as we have seen, the search for atoms leads to the absence of connection between the parts: things fall apart, the system does not hold together.

Another view that predated and in some ways inaugurated the new paradigm is the *synchronicity* theory of Carl Jung. Jung's ideas are still familiar to many people and have had several revivals since he put them forward. He started from a rejection of causality as a sufficient explanation for the link between events that we call coincidental. His reasons were that, because of the success of quantum mechanics, causality is reduced to a statistical basis and does not apply to individual events or pairs of events. Noting the number of coincidences, he suggested an "a-causal principle of synchronicity" that makes things happen in patterns despite a lack of any apparent physical or other causal connection between them. Some of these events were seemingly commonplace, such as finding the same number on a theater ticket as that of the taxicab that took him there, or seeing similarly dressed people passing him in the street. In his book *Synchronicity*,[29] Jung drew on the work of Kammerer, a zoologist who had elaborately noted many instances of related events that were apparently inexplicable by means of a causal link. Some of these were droll: for example, it is irresistible to quote the story of M. de Fortgibu and the plum-pudding, which Kammerer got from the book *The Unknown* by the astronomer Camille Flammarion[30] (cited by Jung):

> *A certain M. Deschamps, when a boy in Orleans, was once given a piece of plum-pudding by a M. de Fortgibu. Ten years later he discovered another plum-pudding in a Paris restaurant, and asked if he could have a piece. It turned out, however, that the plum-pudding was already ordered—by M. de Fortgibu. Many years afterwards M. Deschamps was invited to partake of plum-pudding as a special rarity. While he was eating it, he remarked that the only thing lacking was M. de Fortgibu. At that moment the door opened and an old, old man in the last stages of disorientation walked in: M. de Fortgibu who had got hold of the wrong address and burst in on the party by mistake.*

One smiles, but, as the *Reader's Digest* used to say, life is like that.

Jung embarked on an exploration of the basic connections that he felt existed between events not causally related to one another, in particular, astrology and extrasensory perception. Some of these data have been widely criticized as lacking statistical significance; but then, so do the criticisms. However, it is not for their statistical qualities that they are mentioned here, but for their throwing up those occasional striking instances that are so impressive to the people to whom they happen. The statistical task is formidable and often, as, for instance, in the case of precognitive dreams, rests on so many assumptions that it is as hard to disprove the case for precognition as it is to prove it. Such data are therefore denied the status of facts requiring an explanation even before the discussion starts: that is, they must either be denied or they must be explained as due to mere chance. The explaining away of this sort of awkward fact is an example of special pleading so bald that it could be an object lesson on how not to conduct a scientific inquiry.

A MODEL OF THE WORLD

I have called the alternative worldview that is now emerging the iterative-sequential view. In such a view the world consists of the iteration of rules producing sequences of events. It would be expected that such a system would generate cycles of events of a quasi-periodic nature and patterns of a fractal type. The alternative paradigm, the existing one, which is most generally adopted, is what I have called the random-selection model. In this view the events of the world, while they may be under the control of discoverable laws, are in general unrelated to one another (random) and the patterns that emerge are the result of the selection of some of these random arrangements in preference to others according to some principle of survival. This is the prevailing paradigm that is, for example, applied throughout biological science to account for evolution, but it is also invoked to account for more basic physical phenomena such as the arrangements of pebbles on a beach or the aggregation of crystals in a solution. In its correct place, the random-selection model is appropriate, although there are two things that should make us beware of overapplying it. First, we should keep it clearly in mind there is no such thing as a finite random sequence. Because we can only know finite sequences, this means that we can never prove a sequence of numbers to be random. At best, a sequence of numbers can be termed quasi-random, and if events fit such a sequence, then the events may be referred to as quasi-random also. The second point is that there is good reason to believe that quasi-random sequences of events are much rarer than has been thought, particularly when the events are closely linked in time or space or in some other identifiable dimension.

This need not prevent us from applying the random-selection paradigm under certain circumstances, but the model has been overextended in many fields of biological science, and as a fundamental picture of the world it is inadequate. To assume randomness often leads to a denial of the reality of whole classes of events—they just didn't happen. To ignore a relationship where it exists is as unscientific as to assume one where it does not. Such an attitude of willful ignorance springs, I suggest, from a misapplication of one of the primary principles of science, the dictum of William of Occam: "It is vain to do with more what can be done with fewer." This saying, known as "Occam's Razor," is often quoted in the form "Explanations should not be multiplied unnecessarily," although Occam himself never expressed it in quite those words.[31] Although it does not apply universally,[32] this principle of economy is rightly considered a cornerstone of the scientific method.

What is the application of Occam's razor in these cases of remarkable coincidences? Every time the supposed single explanation—mere coincidence—is applied, it is really a different *ad hoc* explanation that is being invoked, because a lack of systematic relationship can only be asserted if we propose (or assume) another contingent relationship that applies in the particular case. On each occasion the apparent coincidence is explained by the presumption of a different set of explanatory events; this may well overstretch credulity. If, on the other hand, a principle of coincidence is accepted as a basic feature of the world, it would be much more economical to assume that this was the explanation than to look for another. Only if the principle of coincidence failed, would we look further—that is, "multiply explanations."

An interesting sidelight on this issue is the fact that coincidences are often remarked upon for their meaningless nature. This is a comment frequently made by both believers and nonbelievers in the random-selection principle. Some Jungians claim that true synchronicity is very rare, and that meaningless coincidences are pseudo-synchronistic. Only an event that has significance for the life of the person experiencing it can, it is claimed, be called truly synchronistic. The problem with this is that whether something is significant or not is a matter of judgment; this would leave the identification of synchronicity up to the observer.

The iterative-sequential paradigm does not imply that coincidences have any meaning beyond the fact of their occurrence. To suppose that a coincidence must have meaning is to invoke a very different kind of explanation—that of providential happenings, whereby some supervising intelligence is making the coincidences occur. But if coincidences are systematic, they will often be meaningless. They would not be expected to convey meaning beyond that of illustrating by their occurrence the nature of the dynamic and iterative world in which they occur. It would therefore be quite wrong to confuse the iterative-sequential view with a providential one; on the contrary, if events are systematic, they require no special explanation.

SIMPLICITY AND ITERATION

Let us now move to a model of the world based on two principles: simplicity and iteration. We will assume that the initial state of things was a simple one, describable in some way such as a single collection of protonic mass. We will also assume that the observed complexity of its systems is the outcome of the iteration of interactions between their elements. Let us further assume that there are only a few principles of interaction—or perhaps even only one principle, the one-line equation or Theory of Everything sought by the physicists—so the laws governing this initially simple system are themselves simple. With these assumptions, let us see how far it is possible to use this model world to explain the properties of the real world and in particular of complex systems like those we have been considering.

Such a model world displaying iterative properties would be expected to contain certain features. It would have fractal characteristics; that is, it would look similar at different scales of magnification. It would contain complex forms, the complexity of form increasing with increasing iteration number and hence time. Some of these forms would be bounded by highly complex, convoluted curves or surfaces. Because chaotic processes are associated with fractal attractors, iteration of certain kinds would produce chaotic systems. One of the properties of such systems is that they are characterized by parameters that display constancy, particularly parameters of scaling, giving rise to self-similarity. A chaotic system would also lead to the emergence of attractors, and these would correspond to variables that follow quasi-periodic orbits. This model world would have a lower bound in time, with initial conditions, but it would not have an obvious final state. Among the overall characteristics of such a world would be the iterative process itself, which would define a fundamental cycle time for every process in the world. As these cycles continued, more and more information would be added to the system, increasing its apparent entropy, while existing system information was continuously annihilated, leaving total entropy unaltered.

It can be seen from this summary that the overall characteristics that would be expected in such an iterative model world are very close to those known to prevail in the real world. We live in a world dominated by the appearance of the same forms at widely different scales: the elliptical orbit of one body around another under a centrally attracting force is one such feature; the fractal structures of biological organisms[33] are another. The world is apparently dominated by parameters having invariance, such as the speed of light and the value of the gravitational constant, while attractors govern many observable dynamical systems, which function in a quasi-periodic way. The world had an apparent start the so called Big Bang and has progressed through units of time, measured by clocks, which, whether cosmological, geological, or man-made, are iterative devices. Increasingly complex forms have emerged over time: for example, condensed matter; at first baryons and atoms, then molecules, and then the long molecules that form the basis of life. At a certain stage more complex forms emerged as the result of fluid dynamical systems and organic life; these have fractal characteristics and express the subtle curves and surface characteristics of advanced organisms. In the structure of these complex forms the role of information seems to be crucial and leads to decreased entropy, viewed from an outside standpoint, while in fact, chaotic destruction of information about the past leaves entropy unaltered.

In all these respects, the real world shows exactly the characteristics one would predict from a model of the kind I am proposing; one that has a few fundamental laws iterated over a comparatively short period of time. In contrast with this, the features of the observable world do not correspond very well with the assumption of the fundamental randomness of events and the consequent need to assume the emergence of order from disorder only as a result of selection randomly applied over very long time periods.

FINDING YOUR WAY IN A FRACTAL WORLD

I sometimes think of writing—maybe someone has already done it—a book called *The Hitch Hiker's Guide to the Mandelbrot Set.* If you travel about in the set,[34] zooming in and out, enlarging and contracting the display, and moving from place to place, you often find that you want to get to places you have visited before but you can't find your way back to. It is easier if you keep accurate records of your coordinates, noting the position each time you come across an interesting feature, such as the valley of the sea-horses or the desert of the far tip.

As you go around the posterior of the set, into its infold, you can find whorls that contain increasing integral numbers of filaments. These progress in simple arithmetic series. This world is most intriguing: all the integers are to be found in succession as you go deeper into the cleft of the set. Such a world that manufactures all the natural numbers is highly suggestive, given that from the natural numbers, the rest of mathematics can be deduced. Around the main circular trunk of the set there are lands dominated by branches having integral values that do not fall into an arithmetic progression.

This world is not homogeneous but lumpy. Why is this so? It is lumpy partly because its fractal nature puts similar objects next to one another. If you are looking for something, the best place to look is where you are; the chances are that it's right beside you. If you are trying to find new parts of the set, with new features, you do not want to find your way but to lose your way. If you zoom in on a part expecting to see a new pattern, it may well be that you will achieve only a transition to another very similar looking area. In the Mandelbrot set it is easy to go to where you already are; the hard thing to do is to go somewhere new.

Another, connected problem is that of finding your way around in a self-similar world. If we ask "Where am I?" it is not easy to give an absolute answer, only a relative one. The schoolboy who writes in his book: "W. Brown, Form 1A, Hadley High School, Middlesex, England, Europe, The World, The Solar System, The Galaxy" has already sensed this problem and cannot find an answer, so he puts his full address as nearly as he can work it out.

However, such addresses are still relative, with each term giving a location with respect to a larger frame of reference. Where does an address end—that is, become absolute rather than relative? In England we do not put "England" as the last line of the address, because the postman will not go abroad. But for a universal postman, should such a being exist, what address will allow him to find the right letter-box? This problem arises on the Internet, which has universal resource locators (or URLs), which are addresses in global terms. But this does not tell us where the globe is. There is still no absolute address. We can apparently only locate ourselves with relation to other objects in the space in which we find ourselves.

IT'S A COMPLEX WORLD

As we have seen, it is characteristic of living systems that they manufacture complexity. The problem is to limit the level of complexity so that it does not become excessive and uncontrolled. Complexity is now a study in its own right: societies, journals, and institutes[35] are devoted to complexity studies. Many problems in contemporary life could be described as originating in or arising out of complexity. We live in a highly complex world and all the time it is getting more complex. One problem is the sheer growth in human numbers; every year 90 million more people are added to the population of the earth. Although the numbers in westernized societies are temporarily static, no one knows how long this will last. Historically, the trend has always been toward increase.

Another problem, or rather set of problems, is the ever-increasing complexity of technological progress, which means that people constantly have to acquire new skills. Again, historically, there has been a continual and accelerating advance in technology, multiplying the number and type of devices we are expected to cope with. As science becomes more complex, so too do the number of facts and skills that must be learned before education is complete.

There is nothing new in complexity, of course. Even in the simplest social organizations there is a web of interactions and communications, a host of obligations and conventions, as well as the rules of social behavior, that the developing child must absorb. One difference between present and former times is that these processes are now open to study and analysis. At one time the process of learning such communication was unconscious, but now we are much more aware of how we are learning, and why and what we are saying, and how we are saying it. Freud, Marx, and Weber, whether we agree or disagree with them, have made us aware. University departments are dedicated to the study of society and its functioning, of communication and its effects, and of organization and its science. We have become more self-conscious. The media spend a good deal of their time in examining themselves. Organizations are nowadays much addicted to mission statements and vision statements. The time spent by society in navel-contemplation has grown and grown until everyone is almost continuously conscious of what they are doing. Enhanced self-consciousness is not always helpful in accomplishing a complex task; running downstairs, giving a speech, or making love are just three obvious examples.

It might seem perverse to try to introduce yet another metaanalysis—a science of complexity. But there is no simple way out of complexity: the real hope for controlling complexity is to understand it. But the most compelling reason for studying complexity is that it seems to lie at the heart of the unexplained phenomenon of life. Is life the outcome of organized complexity, and if so, how does it work? How can many simple parts come together to form a whole that is something more than any of them separately? This is the problem of emergent properties: a system of simple parts, simply connected together, may often display very startling and unexpected behavior. In a suitable system such as a neural network this can even be behavior of a quasi-intelligent kind. The inverse of complexity is simplicity, and systems that appear complex are often, because of their reducibility, best tackled in terms of a few principles rather than many.[36] It is as though pattern emerged from complexity simply by virtue of the number of elements in the system and their ability to organize themselves into patterns.

THE REALITY OF TIME

It is a principal vice of contemporary Western society that we do not recognize the nature of time. Having cut ourselves off from all natural long-term markers of the passage of time and come to rely on the digital rather than the analog clock, we almost manage to convince ourselves that there is no long-term secular change going on, and that only a moment-by-moment time exists. The digital clock bears more relation to the balance sheet than it does to the rotation of the earth: time of this digitized sort has become a kind of money.

Westerners fear time; they see it as a deadly enemy that will destroy them. On the one hand they try to behave as if time did not exist, literally taking no thought for either tomorrow or yesterday; on the other hand they see time as moving them irresistibly toward their doom. In fact, the one view is a consequence of the other. Because it is seen as their greatest enemy, modern humans fight to abolish time. But this enmity of time is an illusion: how can time destroy any more than space can destroy? On the contrary, time and space are the limits of existence and both space and time are essential to the existence of things in the world. Time cannot destroy, it can only complete.

The intolerance of time leads to the attempted abolition of time. The entertainment, or rather, the distraction, of humans is an important aim of modern life. But time can no more be abolished than space can. Life must exist in time and in space and we must judge and perceive it as a whole.

The denial of the reality of time is part of the generalized anthropocentric attitude by which we seem to be at the center of everything. We feel that our viewpoint is physically central in the world, and this illusion of being physically privileged springs from the fact that our viewpoint is psychologically privileged: it is where our consciousness seems to be located. This leads us to arrange every event in relation to the self and to ignore what is outside our immediate consciousness. In terms of space this is not quite possible, because the boundary of what we can see is relatively distant; but our experience of time excludes the direct perception of events further than about a second from the present. People often say about time: "The past and the future have no reality; there is only now." It would be very odd if someone were to hold up a ruler and say: "The part of the ruler I am holding in my hand exists, but the other parts, to the left and the right of that part, do not exist. There is only here." Such an idea applied to space would be thought mad, yet if it is applied to time, it will usually win at least a momentary consideration.

The difficulty of studying our own consciousness is the difficulty of how to cease to use it and observe it instead. Since we experience consciousness recursively, we find it hard to view it as iterative This recursive nature of consciousness is the reason why people think of time as meaningless "There's only now." They cannot see time as the serial process it is, just because they cannot see their own recursive view of things. It is this readiness to deny the nature of time that is one of the biggest limitations on man's cognitive powers. A perception of the reality of time would extend our cognitive horizons in much the same way that the Copernican insight that the earth was not at the center of the universe expanded our idea of space.

REASON AND MORALITY

Although people pay great respect to the notions of rationality and morality, they often behave as though they possessed neither. If the behavior of someone at a reasoning task is analyzed, the outcome of their reflections is often exactly the same as if they had not thought at all, but followed a simple rule of thumb—a so-called heuristic—that saved them the effort of thinking. Similarly, although people often parade their moral principles as though they were of great importance to the conduct of their lives, in practice the decision they make in a moral dilemma is often as much determined by self-interest as anything else. A whole philosophical outlook—cynicism—is founded on these kinds of observations, but to become cynical in everyday life would be to take too simple a view. Sometimes people do reason rather than just reacting, and they sometimes do follow their principles against their immediate self-interest. Perhaps the reason for the existence of these faculties is for just these rare occasions, choice points, where a decision is called for and where it will influence the future course of events—the course of a life, say, or the future of a nation—and that is their raison d'être.

Chomsky was for the most part wrong and Skinner for the most part right: most of the time people use language in a habitual rather than a thoughtful way. But when a new phrase is coined or some figure of speech is conceived, then structure must be considered and thought becomes necessary. It is for these moments—frequent in the life of a poet, rare in the lives of the rest of us—that the capacity for creating language rather than the habitual response of producing it is found. And what is true of language is perhaps true also of other faculties. Humans are free to choose, although we rarely do so. For the most part we live blindly, following heuristics while mouthing platitudes. But every now and then we can raise ourselves up and make a decision, and it is for these few moments—truly momentous—that the faculty exists. The moment when the ape sees that one stick may be used to reach a longer and more useful one; the moment in someone's life when they decide to take a stand on some issue—it is for these that the capacity for moral choice and rational thought are lying dormant in living beings: it is just for these supreme existential moments that the faculties of reason and virtue exist.

THE MEANING OF LIFE?

It seems that the world exists to express the results of calculations—real computations carried out as iterative processes within it. These are real processes, not symbolic ones. No person, no consciousness, interprets or understands them—they are concrete operations. To give them meaning requires a step from the level of description to that of meaning (or semantics, to use John Searle's[37] term). All that human beings seemingly need to do to give meaning to a set of events is to perceive them. This is the "meaning of meaning," itself inevitably a recursive concept. Perhaps if we can only perceive life processes accurately, we can give them meaning too; their meaning will then spring into existence automatically.

People sometimes ask what their own mental events mean-for example, they often want to know the significance of a dream. It must have *meant* something, they think. But it is very rare that they ask what external events mean. Meaning is assumed to apply only to things over which we have control. If we ask, “What is the meaning of life?” one possible answer is that it lies in our perception of the processes that have produced it. The shapes of living things represent the working out of iterative processes in a way that allows the representation of these results—results that could not otherwise be realized. The nature of those results is that they embody apparent contradiction, so if this is the meaning of life, it is still a great mystery.

THE VIRTUES

Where then are the virtues for which we all seem to strive? Not in any static pattern, for stasis is death. In order for us to stay alive, the patterns within us must continually change and shift. Nor can they lie in any codified rules, for behavior must change just as the shape of the living thing changes. Virtue is not always the same; each new occasion demands a different kind of virtue from those who would be good. What is truth if it need not be fixed, but changes its value from one iteration to the next? If everything is change and there are no static values, should we despair, or fall into a postmodern indifference, offering no hope for humankind in its very obvious dilemmas? Should we become cynical and abandon old ways to replace them with nothing? Or should we adopt the idea that good and evil are not only irrelevant but are even themselves a kind of vice—“the sickness of the mind” in the Buddhist phrase?

Only in recursion, I suggest, is there some hope of escape from this moral, esthetic, cognitive abyss that seems to open at our feet. If there are no fixed truth values, there is still the value of Truth. If there is no unchanging commandment or set of commandments for wise behavior, then at all times Wisdom is still to be found. A higher form of good is revealed in the changing nature of these goods; a higher truth in the very pattern of change itself. Beauty can encompass ugliness, the literature of horror can be an esthetic experience, and the contemplation of terrible crimes can lead us to an understanding of goodness.

A RECURSIVE UNIVERSE?

It seems that we cannot imagine life without time. But is there a way in which we might do so? To describe life in iterative terms we say that one event happens after another in sequence. But in recursive terms the same series of events happen together. The essence of the recursion is that something is defined in terms of itself; in recursive definition the whole of life is in a sense always there.

Perhaps the way in which we can see outside time, some form of meditation or contemplation, lies in the basis of consciousness itself. When we are aware of being aware, when we are fully conscious, we ourselves become in a sense recursive. This feeling of self-awareness is at the heart of every mystical tradition in what Aldous Huxley has called "the perennial philosophy."[38] Meditation is a process of self-awareness in which we are aware as far as possible that we *are* aware. Rather than focusing our awareness on outer objects as in normal life, we refocus it on our own inner consciousness. This reflexive process is the essence of meditative practice and it brings an experience of a different kind from that of the subject–object relationship of everyday life. But such moments are rare, and particularly so in modern life, which is full of distraction and continually draws our attention to outer objects rather than to the fact that we are conscious—the fact of our own awareness. There is little or no opportunity for such an experience in the presence of televisions, videos, computers, and electronic games.

We live in an iterative universe and sometimes we are aware of feeling trapped by it. As Heraclitus put it, we step and do not step into the same rivers, but if we can step outside the iteration and view it instead as a recursion, which is what the practice of meditation is designed to do, then our experience becomes recursive and in a sense we stand outside time, and can look at things as a whole rather than as a succession of moments.

The answer people sometimes make to this is that the meditator is still inside time. After all, we can look at them and see time passing as they sit with eyes closed or open and apparently gazing vacantly. But this is not so; it is we who are inside time and can sense it passing, and we regard the meditating one as yet another object in time. The meditator is outside time.

A MOMENT OF TIME

A moment of time is hell; a moment of eternity is heaven. A moment is not necessarily something in time; it is a unit of experience, not of the world. When we iterate, when we become conscious that we live through time as we are condemned to do, we experience the hell of simply existing. When we become conscious that we exist, that is to say, that we exist eternally, we experience heaven directly.

If we can describe the world as an iteration, with time as an iterative process, can we also define the universe as a recursion, and can a timeless view of the universe be defined recursively? Classically the definition of the universe is a quality of God—that is, the ground of our being, which also has the property of existing outside the observable (but not necessarily knowable) universe. It would also follow that we have a recursive existence corresponding to our iterative existence, and that this too stands outside time, and outside the observable universe.

In Plato's writings the world of appearances is seen as a projection from a higher dimension, as in his analogy of the prisoners chained within a cave, watching the shadows of outside events flickering on the wall.[39] Certainly as the Platonist Socrates went to his death, he believed in the indestructibility of his soul, as something that extended outside the world of time and change.

GOD

What is the nature of God?

It must be almighty; it must be the law-giver, the prescriber of laws.

It must lie outside the universe; pantheism, the belief in the world as God, is a mere limitation of worship to the visible, which enobles nothing. Yet it must be evident or immanent in the universe; it must be a principle that is seen to be working here in this world.

Does it decide from minute to minute the fate of individual men? Does prayer work by bending the ear of the almighty? No; that is a nursery god who sends us presents. It is only by putting ourselves into some kind of resonance with the principle that governs all, the logos, that we can perceive the meaning of the term *God*.

That which gives laws, which prescribes the system, that which lies outside the world: is that God? What is God cannot exist within its own terms, for it contradicts itself in many ways. Its goodness created evil, its power cannot limit itself. God exists and does not exist: God both does and does not exist, bringing moments of awful doubt to the atheist as well as the believer. God is and is not: God is the ultimate contradiction, that is to say, the logos. In a sense it is already there, but outside time; God awaits us, yet though always present, we cannot fully contact it. From beyond even the future it writes, and does not write, the recursive formula for the iterative present. Only in rare moments of enlightenment do we step outside time and see things as they really are.

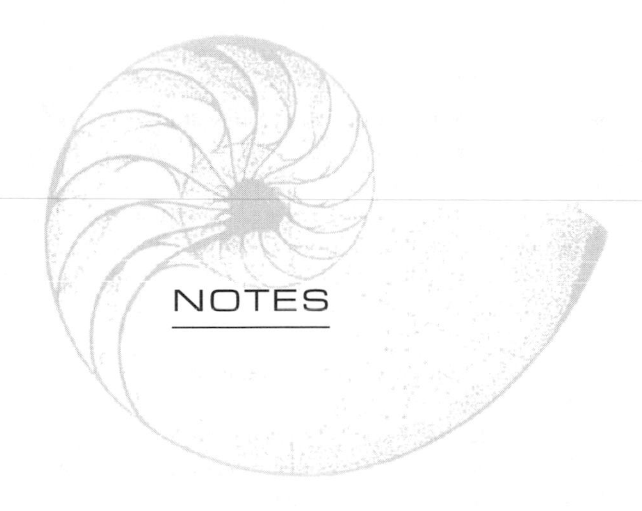

NOTES

CHAPTER 1

1. G. S. Kirk, *Heraclitus: The Cosmic Fragments* (Cambridge, England: Cambridge University Press, 1954), Group 11, Fragment 12. All quotations in the text with the Fragment number noted are from this source.
2. Another example of a chaotic iteration is the trigonometric relation sin x. If you iterate this on a scientific calculator by pressing the *sin* button when you are in radian mode, you can see the values dot around in a chaotic fashion, because *sin* is a nonlinear function.
3. That is, if the values of one of the system variables (such as velocity) are plotted against time on a graph, they will fall, not on a straight line, but on a curve.
4. It is interesting to ask why equally sudden rises in prices, or recoveries, do not happen as frequently as crashes. One reason seems to be that people take fright more easily than they become hopeful.
5. B. Mandelbrot, *The Fractal Geometry of Nature* (New York: Freeman, 1982).
6. In fact, the same problems would arise if you asked for a definition of length or height, but because these are perceptually obvious in a way that time is not, we do not bother to think about it. If asked to define distance, someone might answer that it was the way we measured the arrangement of objects. This definition is just as circular as the one suggested for time.
7. A femtosecond is a trillionth of a second, i.e., 1/1,000,000,000,000 sec.
8. One time-measuring device that is not a clock is a sundial, which might seem at first to contradict this. However, to tell the time we must subdivide the angle moved by the shadow into hours, and this again is an iterative process.
9. For example, if we want to find the value of a factorial number, we do it by multiplying 1 successively by 2, 3, 4, and so on. Factorial n (written $n!$) is defined as

$n! = (1 \times 2 \times 3 \times 4 \times 5 \times \cdots \times n)$. Starting at 1, the successive values generated by this iterative process are 1, 2, 6, 24, 120, etc. The first iterate is 1, the second iterate is 2, etc.

10. A new arrangement of a set of objects is known in math as a permutation. For example, there are six permutations of the symbols in the set {1,2,3}. They are {1,2,3}; {1,3,2}; {2,1,3}; {2,3,1}; {3,1,2}; {3,2,1}. I will use term sequence to imply the possibility of such a reordering.
11. Factorial 51, which is about 10^{66}.
12. For example, a card in position 2 of the deck will move successively through positions 3, 5, 9, 17, 7, 14, 27, and 2.
13. Twist one face at 45 degrees and pry out the middle edge-piece of this rotated face with a blunt knife blade. The other pieces then come out easily.
14. Morwen P. Thistlethwaite, quoted in D. Singmaster, *Notes on Rubik's Magic Cube,* 5th ed. (London: Polytechnic of the South Bank, 1981; published by the author).
15. We once had a pack of cards that got into such a sequence that in the end we decided to throw it away. Each game we played (we played cribbage) came out badly, in the sense that it was a runaway victory for one person, which makes for a boring game. Shuffle the cards as we might, we could not get rid of the "bad karma" that they had somehow acquired. Of course, all we had to do was arrange them into the order they were in when we bought them. But the act of arranging them would simply have meant that our brains would have absorbed the sequence (i.e., the bad karma) lost from the cards! This was, of course, a joke we had with each other and with the cards. The serious point is that sequences once acquired tend to remain.
16. D. Rothstein, E. Henry, and J, P, Gollub,. "Persistent patterns in transient chaotic fluid mixing," *Nature* 401 (1999): 770–772.
17. J. L. Cornell, *Experiments with Mixtures* (New York: Wiley Interscience, 1990).
18. A. J. Wakefield, S. H. Murch, A. Anthony, et al., "Ileal-Lymphoid-Nodular Hyperplasia, Non-Specific Colitis, and Pervasive Developmental Disorder in Children," *Lancet,* 351 (February 28, 1998): 637–641; H. Peltola, A. Patja, P. Leinikki, et al., "No Evidence for Measles, Mumps, and Rubella Vaccine-Associated Inflammatory Bowel Disease or Autism in a 14-Year Prospective Study," *Lancet,* 351 (1998): 1327–1328.
19. K. Töllner, R. PopovitzBiro, M. Lahav, and D. Milstein, "Impact of Molecular Order in Langmuir–Blodgett Films on Catalysis," *Science,* 278 (1997): 2100–2102.
20. This story has been well told by Philip Ball in *Life's Matrix: A Biography of Water* (London: Weidenfeld & Nicolson, 1999).
21. It is unavailable in English translation at the moment. A French edition is Jean-Yves Guillaumin, Translator, *Institution Arithmétique* (Paris: Les Belles Lettres, 1995).
22. There are, for example, some numbers that are squares and this means that they

can be represented as a square array of unary digits. Similarly, some numbers are triangular, others hexagonal, and so on.

23. Such practices at least have not changed!
24. D. Howell, "The Shape and Meaning of the Old English Poem 'Durham,' " In D. Rollason, M. Harvey, and M. Prestwich, Editors, *Anglo-Norman Durham* (Woodridge: The Boydell Press, 1994).
25. The golden mean is the division of a line into two parts, so that the ratio of the smaller part to the larger is equal to the ratio of the larger to the whole, approximately 1:1.62. The golden ratio is the same proportion when used in a painting or drawing and was the approximate ratio of the dimensions of early cinema screens. Television and computer screens have a ratio of more like 1:1.35.
26. D. Hume, "Of Miracles," Chapter 10 of *An Enquiry Concerning Human Understanding* (London: Oxford University Press, 1955).

CHAPTER 2

1. M. G. H. Coles,"Modern Mind-Brain Reading—Psychophysiology, Physiology, and Cognition," Presidential Address to the SPR; reprinted in *Psychophysiology* 26 (1989): 251–269.
2. M. S. Gazzaniga, "How to Change the University," *Science* 282 (1998): 237.
3. The emergon is not, of course, meant to be thought of as an atom in the sense of being a particle. It is a construct for describing the science in question (as atoms sometimes turn out to be).
4. What are sometimes called chemical equations are not equations in a mathematical sense—they are statements about reactions between molecules.
5. B. D. Strahl and C. D. Allis, "The Language of Covalent Histone Modifications," *Nature* 403 (2000): 41–45.
6. J. A. Downie and M. Parniske, "Fixation with Regulation," *Nature* 420 (2002): 369–370.
7. Such studies are now beginning to emerge, e.g., H. Kitano, "Systems Biology; A Brief Overview," *Science* 295 (2002): 1662–1664.
8. P. E. Ahlberg, "Coelacanth Fins and Evolution," *Nature*, 258 (1992): 459.
9. J. H. Schwartz, *Sudden Origins: Fossils, Genes, and the Emergence of Species* (New York: Wiley, 1999).
10. Recently a 21st amino acid has been identified, increasing the time required even more.
11. Attributed to Thomas Huxley.
12. R. Dawkins, *The Blind Watchmaker* (London: Penguin Books, 1988).
13. D. E. Nilsson and S. Pelger, "A Pessimistic Estimate of the Time Required for an Eye to Evolve," *Proceedings of The Royal Society of London Series B—Biological Sciences* 256 (1994): 53–58.

14. R. Dawkins, *Climbing Mount Improbable* (London: Viking, 1996).
15. This is particularly true in view of the very short time scale of some of the "textbook" examples of evolution discussed in the next chapter.
16. G. A. K. Marshall, "On Diaposemitism, with Reference to Some Limitations of the Mullerian Hypothesis of Mimicry," *Transactions of the Entomological Society* (1908): 93–142.
17. R. A. Fisher, *The Genetical Theory of Natural Selection* (London: Oxford University Press, 1930).
18. R. Dawkins, *The Selfish Gene*, 2nd ed. (Oxford and New York: Oxford University Press, 1989).
19. D. B. Ritland, "Unpalatability of Viceroy Butterflies (*Limenitis archippus*) and Their Purported Mimicry Models, Florida Queens (*Danaus gilippus*)," *Oecologia* 88 (1991): 102–108.
20. J. Mallet and M. Joron, "Evolution of Diversity in Warning Color and Mimicry: Polymorphisms, Shifting Balance, and Speciation," *Annual Review of Ecology and Systematics* 30 (1999): 201–233.
21. J. Mallet, "Causes and Consequences of a Lack of Coevolution in Mullerian Mimicry," *Evolutionary Ecology* 13 (1999): 777–806.
22. M. P. Speed, "Robot Predators in Virtual Ecologies: The Importance of Memory in Mimicry Studies," *Animal Behaviour* 57 (1999): 203–213.
23. D. D. Kapan, "Three-Butterfly System Provides a Field-Test of Müllerian Mimicry," *Nature* 409 (2001): 338–340.
24. J. C. Willis, *Age and Area. A Study in Geographical Distribution and Origin of Species.* With Chapters by Hugo de Vries [(and others)] (Cambridge, England: Cambridge University Press: 1922).
25. J. C. Willis, *The Course of Evolution by Differentiation or Divergent Mutation Rather Than by Selection* (Cambridge, England: Cambridge University Press, 1940).
26. D'Arcy W. Thompson, *On Growth and Form* (Cambridge, England: Cambridge University Press, 1942).
27. This will be explored in more detail in chapter 6.
28. J. Cohen and I. Stewart, "The Information in Your Hand," *Mathematical Intelligencer* 13 (1991): 12–15.
29. If we transcribe the transitions from one base to another for the two strands separately, we can produce two shapes in a three-dimensional space—and these shapes are mirror images (enantiomorphs) of one another. See A. J. Koch and J. Lehmann, "About a Symmetry of the Genetic Code," *Journal of Theoretical Biology* 189 (1997): 171–174.
30. R. L. Webber and H. Blum, "Angular Invariant in Developing Human Mandibles," *Science* 206 (1979): 689–691.
31. C. V. Pepicelli, P. M. Lewis, and A. P. McMahon, "Sonic Hedgehog Regulates Branching Morphogenesis in the Mammalian Lung," *Current Biology* 8 (1998): 1083–1086.
32. G. B. West, J. H. Brown, and B. J. Enquist, "The Fourth Dimension of Life: Fractal

Geometry and Allometric Scaling of Organisms," *Science* 284 (1999): 1677–1679.

CHAPTER 3

1. See chapter 2 for a definition of the emergon.
2. See chapter 7 for more on this assertion.
3. C. Darwin, *The Origin of Species* [1859] (London: Penguin Books, 1968), p. 202.
4. Though it may alter gene expression.
5. J. C. Willis, *The Course of Evolution by Differentiation or Divergent Mutation Rather Than by Selection* (Cambridge, England: Cambridge University Press, 1940).
6. D. W. Rudge, "Taking the Peppered Moth with a Grain of Salt," *Biology & Philosophy* 14 (1999): 9–37; J. B. Hagen, "Retelling Experiments: H. B. D. Kettlewell's Studies of Industrial Melanism in Peppered Moths," *Biology & Philosophy* 16 (1999): 39–54.
7. B. S. Grant, D. F. Owen, and C. A. Clarke, "Parallel Rise and Fall of Melanic Peppered Moths in America and Britain," *Journal of Heredity* 87 (1996): 351–357.
8. S. J. Gould and N. Eldredge, "Punctuated Equilibrium Comes of Age," *Nature* 366 (1993): 223–227.
9. M. Barrier, R. H. Robichaux, and M. D. Purugganan, "Accelerated Regulatory Gene Evolution in an Adaptive Radiation," *Proceedings of the National Academy of Sciences of the United States of America* 98 (2001): 10208–10213.
10. S. W. Ulam, "How to Formulate Mathematically Problems of Rate of Evolution?" in P. S. Moorhead and M. M. Kaplan, eds., *Mathematical Challenges to the Neo-Darwinian Interpretation of Evolution* (Philadelphia: Wistar Institute Press, 1967).
11. S. Kauffman, *The Origins of Order* (New York and Oxford: Oxford University Press, 1993).
12. The Game of Life is a cellular automation played out in a grid of light- and dark-colored squares. From three simple rules, exotic patterns can be developed, though most games terminate after only a few moves. For some recent work, see S. Ninagawa, M. Yoneda, and S. Hirose, "1/f Fluctuation in the 'Game of Life,'" *Physica D* 118 (1998): 49–52

CHAPTER 4

1. κυκεον A drink made of ground barley, grated cheese, and wine. If the drink was not stirred, the barley and cheese would settle out.
2. The name *calculus* is derived from "little stone," and from ancient times stones were used as counters. Newton did not call his method by that name; it is a later addition.
3. The Newton–Raphson method for solving equations is to take an approximate

value for a root of the equation and insert it into the equation, and then repeat the procedure with the new value of the root obtained, and so on.

4. The debate about infintesimals is connected with the question whether the ultimate nature of things is discrete (divided into tiny units) or whether it is continuous and the graininess of things is merely a product of our way of observing and thinking about them.
5. B. Russell, *Mysticism and Logic, and Other Essays* [1917] (London: Penguin, 1953).
6. The three equations are: $\dot{X} = \sigma X + \sigma Y$; $\dot{Y} = rX - Y - XZ$; $\dot{Z} = XY - bZ$.

   When these equations are iterated on a computer with appropriate values of *s*, *r*, and *b* and plotted in three dimensions, they produce the well-known butterfly attractor.
7. According to Lorenz's own account, he truncated seven places of decimals to three.
8. A good example of this kind of relationship is the logistic equation, which is quite a good model of population changes in successive generations of animals in an environment with a limited food supply. Providing the rate of reproduction is high enough, the population varies in a chaotic fashion from each generation to the next.
9. An ill-conditioned equation is one whose solution involves dividing by very small quantities, a method that yields increasingly inaccurate results as these quantities get near to zero.
10. For an exploration of this theme see J. Cohen and I. Stewart, *The Collapse of Chaos* (New York: Penguin, 1994).
11. This was the title of a book by P. Cvitanovic, *Universality in Chaos* (Bristol, England: Adam Hilger, 1984), but he meant something rather different; i.e., the universal scaling of the bifurcations in the logistic equation.
12. Velocity is speed with direction added, so it varies between plus and minus.
13. A phase space is a space of one or more dimensions constructed out of variables of the system. For example, in the case of the pendulum, it can be made out of the variables of position and velocity, but we could equally use other variables, such as angular velocity and time.
14. P. Totterdell, R. B. Briner, B. Parkinson, and S. Reynolds, "Fingerprinting Time Series: Dynamic Patterns in Self-Report and Performance Measures Uncovered by a Graphical Nonlinear Method" *British Journal of Psychology*, 87 (1996): 43–60.
15. A. Babloyantz and A. Destexhe, "Is the Normal Heart a Periodic Oscillator?" *Biological Cybernetics* 58 (1987): 203–211.
16. Euclidean figures are those treated by Euclid in his books on geometry and include one-dimensional figures (straight lines), two-dimensional ones (circles, squares, etc.), and in three dimensions, spheres or parallelepipeds (boxes)—nothing too complicated. Euclidean figures can be measured with a ruler and their length, surface area, or volume can be calculated. Not so with fractals!
17. For regular fractals, dimension depends on the scaling-down ratio of the replacement process and the number of parts being replaced. We calculate the fractal dimension, *D*, as

$$D = \frac{\log(N)}{\log\left(\frac{1}{r}\right)}$$

where r = scaling-down ratio and N = number of replacement parts. For example, to calculate the fractal dimension of the Koch (snowflake) curve,

is replaced by

So the number of new units is 4, the scaling-down factor is 1/3, and D can be calculated as

$$D = \frac{\log 4}{\log\left(\frac{1}{1/3}\right)} = \frac{\log 4}{\log 3} = 1.261859$$

CHAPTER 5

1. D. Jones, "The Stability of the Bicycle," *Physics Today* (April 1970): 34–40.
2. Homoiothermic (warm-blooded) creatures are those whose blood is maintained in a certain temperature range); poikilothermic (cold-blooded) creatures have the same temperature as their environment
3. Such a series of values of one variable is called a time series. Time-series analysis is a large part of the application of chaos theory to behavioral and physiological data.
4. F. Takens, *Dynamical Systems and Turbulence* (Lecture Notes in Mathematics) (Berlin and New York: Springer Verlag, 1981).
5. This is true for the majority of chaotic systems, although there are boundary cases that are exceptions.
6. A. Goldberger and B. J. West, "Applications of Nonlinear Dynamics to Clinical Cardiology," *Annals of the New York Academy of Sciences* 504 (1987): 195–213.
7. N. M. Magid, G. J. Martin, and R. F. Kehoe, "Diminished Heart-Rate Variability in Sudden Cardiac Death," *Circulation* 72 (1985): 241.
8. A fighter airplane that flew like a bird would indeed be a stealth weapon, and would have an unlimited range because it would use very little fuel!
9. K. P. Dial, A. A. Biewener, B. W. Tobalske, and D. R. Warrick, "Mechanical Power Output of Bird Flight," *Nature* 390 (1997): 67–70.
10. L. Sirovich and S. Karlsson, "Turbulent Drag Reduction by Passive Mechanisms," *Nature* 388 (1997): 753–755.
11. Compare the much quoted "Life is complexity at the edge of chaos."

12. We are so used to equating the two that this may seem a nonsensical statement; but IQ scores are only a way of measuring intelligence, not intelligence itself.
13. Zoic rocks are those bearing fossils.
14. A. V. Holden, J. Hyde, and H. Zhang, "Computing with the Unpredictable; Chaotic Dynamics and Fractal Structures in the Brain," *Proceedings of the British Computer Society Conference on Chaos and Fractals*, London, England (1992).
15. A. Babloyantz and A. Destexhe, "Low-Dimensional Chaos in an Instance of Epilepsy," *Proceedings of the National Academy of Sciences, USA* 83 (1986): 3513–3517.
16. W. Freeman, "Simulation of Chaotic EEG Patterns with a Dynamic Model of the Olfactory System," *Biological Cybernetics* 56 (1987): 139–150.
17. R. Azouz and C. M. Gray, "Dynamic Spike Threshold Reveals a Mechanism for Synaptic Coincidence Detection in Cortical Neurons in Vivo," *Proceedings of The National Academy of Sciences of the United States of America* 97 (2000): 8110–8115.
18. T. Shinbrott, E. Ott, C. Grebogi, and J. A. Yorke, "Using Chaos to Direct Trajectories to Targets," *Physical Review Letters* 65 (1990): 3215–3218.
19. R. Gregory, *Eye and Brain* (London: Weidenfeld & Nicolson, 1977).
20. K. Wiesenfeld and F. Moss, "Stochastic Resonance and the Benefits of Noise: From Ice Ages to Crayfish and SQUIDs," *Nature* 373 (1995): 33–36.
21. B. Mandelbrot, *The Fractal Geometry of Nature* (New York: W. H. Freeman, 1982).
22. D. L. Gilden, M. A. Schmuckler, and K. Clayton, "The Perception of Natural Contour," *Psychological Review* 100 (1993): 460–478.
23. A. Garfinkel, W. L. Ditto, M. L. Spano, and J. N. Weiss, "Control of Cardiac Chaos," *Science* 257 (1992): 1230–1235.
24. L. Lipsitz, and A. L. Goldberger, "Loss of 'Complexity' and Aging: Potential Applications of Fractals and Chaos Theory to Senescence," *Journal of the American Medical Association* 267 (1992): 1806–1809.
25. L. Glass, M. R. Guevara, J. Bélair, and A. Shrier, *Physical Review, A* 29 (1984): 1384–1357.
26. L. Glass and M. C. Mackey, *From Clocks to Chaos. The Rhythms of Life* (Princeton: Princeton University Press, 1988).
27. These critical speeds are eigenvalues of the dynamical equations determining the motion of the strap.
28. *Independent,* London: March 17 (1996).
29. *Sunday Times,* London: May 12 (1996).
30. One major difference between markets and many other systems is that markets are not constrained in the same way.
31. W. Brock, "Distinguishing Random and Deterministic Systems," *Journal of Economic Theory* 40 (1986): 168–195.
32. M. Small and C. K. Tse, "Applying the Method of Surrogate Data to Cyclic Time Series," *Physica D* 164 (2002): 187–201.
33. D. H. Gonsalves, R. D. Neilson, and A. D. S. Barr, "A Study of the Response of a Discontinuously Nonlinear Rotor System," *Nonlinear Dynamics* 7 (1995): 415.

34. T. A. Bass, *The Newtonian Casino* (London: Penguin, 1991).
35. An ergodic orbit is one that, in time, visits all available parts of its phase space.
36. P. A. Scholes, *Oxford Companion to Music*, 10th ed. (London, New York, and Toronto: Oxford University Press, 1970).
37. Spatial frequency corresponds to the way the eye/brain system processes visual stimuli, and is in many ways analogous to sound frequency.
38. W. Dansgaard and others, "Evidence for General Instability of Past Climate From a 250-kyr Ice-Core Record," *Nature* 364 (1993): 218–220.
39. R. B. Alley and others, "Abrupt Increase in Greenland Snow Accumulation at the End of the Younger Dryas Event," *Nature* 362 (1993): 527–529.
40. K. C. Taylor and others, "The Holocene-Younger Dryas Transition Recorded at Summit, Greenland," *Science* 278 (1997): 825–827.
41. Between 250,000 and 10,000 years ago the degree of chaos in the climate, using a chaotic measure called the Lyapunov exponent, was less than 0.6, while after 10,000 years ago it was more than 0.8. [Data from the Greenland Summit Ice Cores CD-ROM, 1997. Available from the National Snow and Ice Data Center, University of Colorado at Boulder, and the World Data Center-A for Paleoclimatology, National Geophysical Data Center, Boulder, Colorado.]
42. C. Nicolis and G. Nicolis, "Is There a Climatic Attractor?" *Nature* 311 (1984):529–532.
43. A. A. Tsonis and J. B. Elsner, "The Weather Attractor Over Very Short Timescales," *Nature* 333 (1988): 545–547.
44. J. Baldock, *The Little Book of Sufi Wisdom* (Shaftsbury, England: Element Books, 1995).
45. Ibn 'Ata 'Illa, quoted in Baldock, reference 44.

CHAPTER 6

1. D'Arcy W. Thompson, *On Growth and Form* (Cambridge, England: Cambridge University Press, 1942).
2. Their Web page is http://www-history.mcs.st-and.ac.uk/history/Miscellaneous/darcy.html
3. Reprinted with permission from F. Grey and J. K. Kjems, "Aggregates Broccoli and Cauliflower," *Physica D* 38 (1989): 154–159.
4. D. Marr and H. K. Nishihara, "Representation and Recognition of the Spatial Organization of Three-Dimensional Shapes," *Proceedings of the Royal Society of London* 200 (1978): 269–294.
5. The set is produced by the iteration $Z' = Z^2 + C$, where Z and C are complex numbers having a two-dimensional interpretation. They are written in capital italics to distinguish them from ordinary numbers, which are one-dimensional.

   The meaning of this iteration is that a variable is squared and another number added to it: this operation is carried out repeatedly, using the new value of the variable in the next iteration. (The iteration is called nonlinear because it involves a higher power than unity in the exponent of the variable—here it is squared. If functions of this type are plotted on a graph, they produce curves rather than straight lines—hence the term nonlinear.)

To make the Mandelbrot set we calculate successive values of the variable for each point in the plane. The value (modulus) of it will either converge, diverge, or oscillate in some way. If it converges to remain smaller than 2, then the point is deemed to be a member of the set; if it diverges to become greater than 2, then it is outside the set. A picture of M as it looks after 150 iterations can be seen in figure 6.8. The black area consists of points that are convergent according to the above definition; i.e., their modulus is less than 2. The rest of the plane is outside the set and is left white. We can express this by saying that linked to iteration 1 is the condition $|Z| \leq 2$. We can thus specify the set M by the following process: for $n = 1$ to 150, $Z' = Z^2 + C$, while $|Z| \leq 2$.

6. There was an excellent April Fool's joke in circulation on the Internet, complete with illustration, alleging that a thirteenth-century Bavarian monk had spent years computing the Mandelbrot set and had incorporated it into an illuminated manuscript so that it appeared as the Star of Bethlehem. Unfortunately, it seems now to have disappeared.
7. D. J. de S. Price, *Gears from the Greeks: The Antikythera Mechanism: A Calendar Computer From ca. 80 BC* (New York: Science History Publications, 1975).
8. F. Petrie, *3000 Decorative Patterns of the Ancient World* (New York: Dover, 1986).
9. For more details on how this striking image, originated by Melinda Green, was created, see her Web site: http://www.superliminal.com/fractals/bbrot/bbrot.htm
10. This seems further to imply the production of such patterns by some spontaneous and unknown computing process going on in the brain of the visionary.
11. I. Shah, *The Way of the Sufi* (Harmondsworth, England: Penguin Books, 1974).
12. G. S. Kren, J. H. Linehan, and C. A. Dawson, "A Fractal Continuum Model of the Pulmonary Arterial Tree," *Journal of Applied Physiology* 72 (1992): 2225–2237.
13. J. G. Fraser, *The Golden Bough* (London: Macmillan, 1922).
14. That is, the points whose modulus satisfies the criterion $|Z| \leq 2$.
15. G. B. West, J. H. Brown, and B. J. Enquist, "A General Model for the Origin of Allometric Scaling Laws in Biology," *Science* 276 (1997): 122–126.
16. J. Stephenson, "Spirals in the Mandelbrot Set I," *Physica A* 205 (1994): 634–664.
17. Jouko Seppanen, personal communication.
18. The fractal dimension of a curve is called its Hausdorff–Besicovich dimension. It can be calculated by working out the size of the curve when measured with a particular size of ruler. In the case of fractals, the length of the curve increases as the ruler-length decreases, as in the case of the coastlines of most countries.
19. L. Wolpert, "Positional Information and the Spatial Pattern of Cellular Differentiation," *Journal of Theoretical Biology* 25 (1969): 1–47.
20. Fujita and others have shown that the budding patterns of yeast (*Saccharomyces cerevisiae*) are influenced by the expression of the gene *AXL1*. This gene is ex-

pressed as a protein similar to the human insulin-degrading enzymes found in humans and *Drosophila* and to the *Escherichia coli ptr* gene product. The normal budding pattern depends upon mating type, and is axial for the haploid **a** or α cells and bipolar for the diploid **a**/α cells. This pattern can be altered, the diploid a/α cells budding axially with the ectopic expression of *AXL1*. This has implications for computed morphology: the pattern of morphological growth will depend on the local ratios of axial and bipolar division, and this in turn depends on gene switching. Axial growth of the diploid **a**/α cells will cease when the expression of *AXL1* ceases. (See A. Fujita and others, "A Yeast Gene Necessary for Bud-Site Selection Encodes a Protein Similar to Insulin-Degrading Enzymes," *Nature* 372 (1994): 567–570.)

21. I. Hori, "Cytological Approach to Morphogenesis in the Planarian Blastema. 1. Cell Behavior During Blastema Formation," *Journal of Submicroscopic Cytology and Pathology* 24 (1992): 75–84.
22. R. J. Bird and F. Hoyle, "On the Shapes of Leaves," *Journal of Morphology* 219 (1994): 225–241.
23. The formula for oak is

$$G(x, y) = -3\,(x + y)\,\{1 - x / 3 - 2|frc[3x(1 - y)] + 0.5|\}$$

The formula for holly is

$$G(x, y) = -x|x|\left[1 + \min\left\{0{,}4y\,\frac{5x + 9}{x + 2}\right\}\right]s_1 s_2 \cdots$$

where $s_1 = 1 + d_1/(|x + 0.25| + \delta_1)$,
$s_2 = 1 + d_2/(|x + 0.8| + \delta_2)$,
$s_3 = 1 + d_3/(|x + y + 0.8| + \delta_3)$,
$s_4 = 1 + d_4/(|x + y + 1.1| + \delta_4)$,
$s_5 = 1 + d_5/(|x + y + 1.4| + \delta_5)$,

and where

$$\delta_1 = \delta_2 = 0.01,\ \delta_3 = \delta_4 = \delta_5 = 0.005,$$

$$d_1 = d_2 = d_3 = 0.03,\ d_4 = d_5 = 0.06.$$

24. G. B. West, J. Brown, and B. J. Enquist, "A General Model for the Structure and Allometry of Plant Vascular Systems," *Nature* 400 (1999): 664–667.
25. B. Sapoval, T. Gobron, and A. Margolina, "Vibrations of Fractal Drums," *Physical Review Letters* 67 (1991): 2974–2977.

CHAPTER 7

1. While there are only 21 amino acids, there are 64 (4 × 4 × 4) possible triplets, so many of them code for the same amino acids.

2. The ability of separated strands of DNA to re-anneal at such a high rate gave an early indication that many segments of DNA must be repeated hundreds of thousands or even millions of times. See E. O. Long and I. B. Dawid, "Repeated Genes in Eukaryotes," *Annual Review of Biochemistry* 49 (1980): 727–764.
3. It has been suggested that "alterations in the redundancy of given DNA sequences and changes in their methylation (inactivation) patterns are ways to produce continuous genotypic variability within the species, which can then be exploited in adaptation to environmental pressures." See A. Cavallini et al., "Nuclear DNA Changes Within *Helianthus annuus L*: Variations in the Amount and Methylation of Repetitive DNA Within Homozygous Progenies," *Theoretical and Applied Genetics* 92 (1996): 285–291.
4. R. Dawkins, *The Selfish Gene* (New York: Oxford University Press, 1978).
5. If there are four possible digits in a system of arithmetic, it is called base-four arithmetic. Similarly, there is base-ten arithmetic, which is the familiar decimal system, and base-two or binary arithmetic used by computers.
6. A. M. Turing, "On Computable Numbers," *Proceedings of the London Mathematical Society, Series* 2 42 (1936): 230–265.
7. R. Benne, J. Vandenburg, J. P. J. Brakenhoff, P. Sloof, J. H. Vanboom, and M. C. Tromp, "Major Transcript of the Frameshifted *COXLL* Gene From Trypanosome Mitochondria Contains 4 Nucleotides That Are Not Encoded in the DNA," *Cell* 46 (1986): 819–826.
8. R. Cattaneo, K. Kaelin, K. Baczko, and M. A. Billeter, "Measles-Virus Editing Provides an Additional Cysteine-Rich Protein," *Cell* 56 (1989): 759–764.
9. These often apparently arise as "errors" but some perhaps are better interpreted as computationally driven mutations.
10. As Jouko Seppanen has pointed out, in a sense DNA can be seen as just such a long-range morphogen, and might have been so called: however, the name "gene" has conferred on DNA a functional fixity that has partially inhibited this perception.
11. W. Driever and C. Nusslein-Volhard, "The Bicoid Protein Determines Position in the Drosophila Embryo," *Cell* 54 (1988): 83–93.
12. B. Houchmandzdeh, E. Wieschaus, and S. Leibler, "Establishment of Developmental Precision and Proportions in the Early *Drosophila* Embryo," *Nature* 415 (2002): 798–802.
13. A. J. Simmonds, G. dos Santos, I. Livne-Bar, and H. M. Krause, "Apical Localization of Wingless Transcripts Is Required for Wingless Signaling," *Cell* 105 (2001): 197–207.
14. G. S. Wilkie and I. Davis, "Drosophila Wingless and Pair-Rule Transcripts Localize Apically by Dynein-Mediated Transport of RNA Particles" *Cell* 105 (2001): 209–219.
15. Y. Chen and A. F. Schier, "The Zebrafish Nodal Signal Squint Functions as a Morphogen," *Nature* 411 (2001): 607–610.
16. K. Nakajima, G. Sena, T. Nawy, and P. N. Benfey, "Intercellular Movement of the Putative Transcription Factor SHR in Root Patterning," *Nature* 413 (2001): 307–311.

17. J. B. Gurdon and P.-Y. Bourillot, "Morphogen Gradient Interpretation," *Nature* 413 (2001): 797–803.
18. R. J. Lipton, "DNA Solution of Hard Computational Problems," *Science* 268 (1995): 542–545.
19. A computationally hard problem is one that takes a very long time—"polynomial time"—to solve, because it has so many possible alternatives. The traveling salesman problem is: What is the shortest route a traveling salesman must take to pass through all the capital cities of the mainland United States? This takes an impossibly long time to solve computationally, because there are 48 cities, giving 48!, which is about 10^{60} possibilities—1 followed by 60 zeroes. There are about 3×10^{16} nanoseconds in a year. To show the computational immensity of the problem, if you could run through the routes at a rate of one every nanosecond, it would take considerably longer than the time the universe has so far existed to solve it algorithmically.
20. M. G. Schueler, A. W. Higgins, M. K. Rudd, K. Gutashaw, and H. F. Willard, "Genomic and Genetic Definition of a Functional Human Centromere," *Science* 294 (2001): 109–115.
21. A. W. Murray, "Creative Blocks: Cell-Cycle Checkpoints and Feedback Controls," *Nature* 359 (1992): 599–604.
22. I. Hori, "Cytological Approach to Morphogenesis in the Planarian Blastema. 1. Cell Behavior During Blastema Formation," *Journal of Submicroscopic Cytology and Pathology* 24 (1992): 75–84.
23. According to the Bird and Hoyle model, a critical factor in this decision may be modeled by the outcome of the computations:

$$\text{Iterate } Z' = g(Z, P) \qquad \text{while } |Z| \leq h(P) \qquad (1)$$

and

$$P' = f(P) \qquad (2)$$

where Z is a complex variable, P is the cell's positional information, and f, g, and h are biocomputed functions. The prime symbol ($'$) represents the new value of the variable calculated each time it is iterated (i.e., during each cell cycle.) The iteration (1) is carried out for each cell, and determines whether it is to divide, maintain itself, or die as a result of the comparison of the modulus of the variable Z (written $|Z|$) with the criterion $h(P)$. The new cell position after division is determined by (2). The interpretation of the functional symbols for the purposes of the iteration is that g is the formula that is being iterated, h is the criterion against which the variable is tested, and f is the function that represents the updating of the positional information when the cell divides.

24. L. Alland, R. Muhle, H. Hou, J. Potes, L. Chin, N. Schreiber-Agus, and R. A. Depinho, "Role for N-CoR and Histone Deacetylase in Sin3-Mediated Transcriptional Repression," *Nature* 387 (1997): 49–55.
25. C. Holden, "The Quest to Reverse Time's Toll," *Science* 295 (2002): 1032–1033.
26. J. Marx, "Tackling Cancer at the Telomeres," *Science* 295 (2002): 2350.

27. S. V. Rajkumar, "Current Status of Thalidomide in the Treatment of Cancer," *Oncology–New York* 15, 67–874.
28. V. M. Weaver and others, "Beta 4 Integrin-Dependent Formation of Polarized Three-Dimensional Architecture Confers Resistance to Apoptosis in Normal and Malignant Mammary Epithelium," *Cancer Cell* 2 (2002): 205–216.
29. J. McClellan and S. Newbury, "Inexpressible Trinucleotides," *Nature* 379 (1996): 396.
30. H. Deissler, A. Behn-Krappa, and W. Doerfler, "Purification of Nuclear Proteins from Human HeLa-Cells That Bind Specifically to the Unstable Tandem Repeat (CGG) in the Human FMR1 Gene, *Journal of Biological Chemistry* 271 (1996): 4327–4334.
31. M. B. Kastan, "Signalling to p53—Where Does It All Start?" *BioEssays* 18 (1996): 617–619.
32. J. S. Steffan and others, "Histone Deacetylase Inhibitors Arrest-Polyglutamine Dependent Neurodegeneration in *Drosophila*," *Nature* 413 (2001): 739–743.
33. J.-H. J. Cha, "Transcriptional Dysregulation in Huntington's Disease," *Transactions in Neuroscience* 23 (2000): 387–392.
34. M. Barinaga, "Forging a Path to Cell Death," *Science* 273 (1996): 735–737.
35. E. Pennisi, "Haeckel's Embryos: Fraud Rediscovered," *Science* 277 (1997): 1434.
36. T. Yokoyama, N. G. Copeland, N. A. Jenkins, C. A. Montgomery, F. F. B. Elder, and P. A. Overbeek, "Reversal of Left-Right Asymmetry—A Situs-Inversus Mutation," *Science* 260 (1993): 679–682; N. A. Brown and A. Lander, "On the Other Hand . . . ,"*Nature* 363 (1993): 303.
37. A. J. S. Klar, "Fibonacci's Flowers," *Nature* 417 (2002): 595.
38. A generalized version of the Turing machine was devised by Turing. The "universal Turing machine" was so called because it could indeed compute any recursively defined function. This was achieved by letting the program reside on the tape rather than in the automaton of the machine itself. This in turn meant that the automaton need have no internal memory, except for its own state and the current symbol, which may need to be written to the tape.

CHAPTER 8

1. N. Eldredge and S. J. Gould, "On Punctuated Equilibria," *Science* 276 (1997): 338–339.
2. S. F. Elena, V. S. Cooper, and R. E. Lenski, "Punctuated Evolution Caused by Selection of Rare Beneficial Mutations," *Science* 272 (1996): 1802–1804.
3. R. E. Lenski and M. Travisano, "Dynamics of Adaptation and Diversification—A 10,000-Generation Experiment with Bacterial Populations," *Proceedings of the National Academy of Sciences of the United States of America* 91 (1994): 6808–6814.
4. R. Leakey and R. Lewin, *Origins Reconsidered: In search of What Makes Us Human* (New York: Anchor, 1993).

5. A. Das and C. D. Gilbert, "Topography of Contextual Modulations Mediated by Short-Range Interactions in Primary Visual Cortex," *Nature* 399 (1999): 655–661.
6. D. I. Perret, K. A. May, and S. Yoshikawa, "Facial Shape and Judgement of Female Attractiveness," *Nature* 368 (1994): 239–242.
7. T. R. Alley and M. R. Cunningham, "Averaged Faces Are Attractive, But Very Attractive Faces Are Not Average," *Psychological Science* 2 (1991): 123–125.
8. G. Rhodes, K. Geddes, L. Jeffery, S. Dziurawiec, and A. Clark, "Are Average and Symmetric Faces Attractive to Infants? Discrimination and Looking Preferences," *Perception* 31 (2002): 315–321.
9. G. S. Marmor and L. A. Zaback, "Mental Rotation by the Blind: Does Mental Rotation Depend on Visual Imagery?" *Journal of Experimental Psychology: Human Perception and Performance* 2 (1976): 515–521.
10. C. Darwin, *The Descent of Man and Selection in Relation to Sex* (London: John Murray, 1913).
11. M. Enquist and A. Arak, "Selection of Exaggerated Male Traits by Female Aesthetic Senses," *Nature* 361 (1993): 446–448.
12. J. A. Heinemann, "Bateson and Peacocks' Tails," *Nature* 363 (1993): 308.
13. W. Bateson, *Materials for the Study of Variation Treated with Especial Regard to Discontinuity in the Origin of Species* (Baltimore: Johns Hopkins University Press, 1992) [first published 1894].
14. M. Ghyka, *The Geometry of Art and Life* (New York: Dover, 1977); H. Z. Huntley, *The Divine Proportion* (New York: Dover, 1970).
15. Song of Sol. 7:2–3.
16. Song of Sol. 5:12–15.
17. K. B. Manchester, *FORUM Collection* 5(2) (1992): 171–172.
18. "Carl": personal communication.
19. H. B. Barlow, "Single Units and Sensations: A Neuron Doctrine for Perceptual Psychology," *Perception* 1 (1972): 371–394.
20. The account of language evolution could be challenged, for example, since stone tools and artifacts have been found that precede modern humans by several hundred thousand years. An alternative hypothesis is gradualism, with a continuum of language use between modern and ancient man. The issue appears very difficult to decide, as it depends largely on the soft tissue of the brain and vocal tract, long dissolved by time.
21. R. L. Cann, M. Stoneking, and A. C. Wilson, "Mitochondrial DNA and Human Evolution, *Nature* 325 (1987): 31–36.
22. M. Balter, "What Made Humans Modern?" *Science* 295 (2002): 1219–1225.
23. A. Gibbons, "Human Evolution—Y Chromosome Shows That Adam Was an African," *Science* (1997): 804–805.
24. J. M. Howard, " 'Mitochondrial Eve,' 'Y Chromosome Adam,' Testosterone, and Human Evolution," *Revista Di Biologia-Biology Forum* 95 (2002): 319–325.
25. E. M. Miller, " 'Out of Africa': Neanderthals and Caucasoids," *Mankind Quarterly* 37, 231–253.

26. P. Mellars, "The Fate of the Neanderthals," *Nature* 395 (1998): 539–540.
27. I. V. Ovchinnikov, A. Gotherstrom, G. P. Romanova, V. M. Kharitonov, K. Liden, and W. Goodwin, "Molecular Analysis of Neanderthal DNA From the Northern Caucasus," *Nature* 404 (2000): 490–493.
28. I. Tattersal, *The Last Neanderthal* (New York: Macmillan, 1995)
29. "Le Viol" (The Rape) 1934.
30. D. Morris, *The Naked Ape* (London: Jonathan Cape, 1967).
31. F. Dostoevsky, *The Brothers Karamazov*, trans. by David Magarshak (Harmondsworth, England: Penguin, 1970), p. 137.

CHAPTER 9

1. This is a verbal definition of a mathematical entity and like all such it is not quite accurate: in mathematics, a random walk is a function p of the real numbers such that

$$p(x) \geq 0$$

$$\sum_{x \in R} p(x) = 1$$

where x is an integer, $p(x)$ is called the *transition function* from 0 to x, and R is the set of reals. This leads to the concept of equal probabilities of motion in any given direction.

2. J. S. Bendat, *Principles and Applications of Random Noise Theory* (New York: Wiley, 1958).
3. An apparent exception is quantum processes, but here again, an underlying order has been proposed. See, for example, D. Bohm, *Wholeness and the Implicate Order* (London: Routledge & Kegan Paul, 1981).
4. Strictly, algorithmic complexity.
5. One such formula for π is the continued fraction

$$\frac{4}{\pi} = 1 + \cfrac{1}{2 + \cfrac{3^2}{2 + \cfrac{5^2}{2 + \ldots}}}$$

This formula, which can be extended indefinitely, contains a comparable number of symbols to the sequence it generates. So we might say that the sequence of digits in π is highly random. However, we could rewrite the formula as

$$\frac{4}{\pi} = 1 + F_\infty ; F_{n+1} = \frac{(2n-1)^2}{2 + F_n}$$

which is a very short program.

6. C. Shannon, *The Mathematical Theory of Communication* (Urbana: University of Illinois Press, 1975).
7. "Awareness Is News of a Difference": G. Bateson, *Mind and Nature: A Necessary Unity* (New York: Dutton, 1979).
8. If the distribution of points in n-dimensional space can be accounted for equally well by assuming a space of $n + 1$ dimensions, the system is said to be chaotic with a correlation dimension of n. If the distribution cannot be accounted for as we go on increasing n, it is said to be random. It may be that by assuming a larger n we could show, if we had enough data at our disposal, that the system was not in fact random but simply chaotic with a high dimensionality.
9. See Arthur Koestler, *The Case of the Midwife Toad* (London: Hutchinson, 1971).
10. T. S. Eliot, "Four Quartets: East Coker," in *The Complete Poems and Plays of T. S. Eliot* (London and Boston: Faber and Faber, 1969).
11. Eliot, "The Hollow Men,"in *ibid.*
12. C. H. Bennett, "Demons, Engines, and the Second Law," *Scientific American* 257 (1987): 108.
13. J. Eggers, "Sand as Maxwell's Demon," *Physical Review Letters* 83 (1999): 5322–5325.
14. K. Wiesenfeld and F. Moss, "Stochastic Resonance and the Benefits of Noise: From Ice Ages to Crayfish and SQUIDs," *Nature* 373 (1995): 33–36.
15. J. J. Collins, T. T. Imhoff, and P. Grigg, "Noise-Enhanced Tactile Sensation," Nature 383 (1996): 770–770.
16. J. E. Molloy, J. E. Burns, J. Kendrickjones, R. T. Tregear, and D. C. S. White, "Movement and Force Produced by a Single Myosin Head," *Nature* 378 (1995): 209–212.
17. P. S. Landa and P. V. E. McClintock, "Changes in the Dynamical Behavior of Nonlinear Systems Induced by Noise," *Physics Reports* (review section of *Physics Letters*) 323 (2000): 1–8.
18. Let a board game have an initial position B(0). Let a move M(1) transform B(0) into B(1) with an operation o:

    $$B(1) = B(0) \text{ o } M(1)$$

    Let a second move M(2) transform B(1) into B(2):

    $$B(2) = B(1) \text{ o } M(2)$$

    A sequence of moves M(1) ⋯ M(n) will transform the board into B(n):

    $$B(n) = B(0) \text{ o } \{M(1)M(2)M(3) \cdots M(n)\}$$

    Since the sequence of moves is certain; that is,

    $$p\{M(1)M(2) \cdots M(n)\} = 1$$

    the entropy of B(n) is the same as that of B(0).

19. V. Serebriakoff, *The Future of Intelligence. Biological and Artificial* (Carnforth: Parthenon, 1987).
20. P. E. Ahlberg, "Coelacanth Fins and Evolution," *Nature* 358 (1992): 459.
21. E. Schroedinger, *What Is Life?* (Cambridge: Cambridge University Press, 1967).
22. In the case of some homeopathic drugs, there can almost certainly be no active chemicals because increasing dilution has reduced the probability of a molecule being present in the dose to almost zero.
23. E. Davenas et al., "Human Basophil Degranulation Triggered by Very Dilute Antiserum Against IgE," *Nature* 333 (1988): 816–818.
24. Samuel Hahnemann, the founder of homeopathy, was a German physician. In 1810 he published *Organon of the Healing Art,* and the following year *Pure Materia Medica.* Hahnemann died in 1843. His method of dilution, in the translation published by Oliver Wendell Holmes in 1842, is as follows:

    A grain of the substance, if it is solid, a drop if it is liquid, is to be added to about a third part of one hundred grains of sugar of milk in an unglazed porcelain capsule which has had the polish removed from the lower part of its cavity by rubbing it with wet sand; they are to be mingled for an instant with a bone or horn spatula, and then rubbed together for six minutes; then the mass is to be scraped together from the mortar and pestle, which is to take four minutes; then to be again rubbed for six minutes. Four minutes are then to be devoted to scraping the powder into a heap, and the second third of the hundred grains of sugar of milk to be added. Then they are to be stirred an instant and rubbed six minutes, again to be scraped together four minutes and forcibly rubbed six; once more scraped together for four minutes, when the last third of the hundred grains of sugar of milk is to be added and mingled by stirring with the spatula; six minutes of forcible rubbing, four of scraping together, and six more (positively the last six) of rubbing, finish this part of the process.

    Every grain of this powder contains the hundredth of a grain of the medicinal substance mingled with the sugar of milk. If, therefore, a grain of the powder just prepared is mingled with another hundred grains of sugar of milk, and the process just described repeated, we shall have a powder of which every grain contains the hundredth of the hundredth, or the ten thousandth part of a grain of the medicinal substance. Repeat the same process with the same quantity of fresh sugar of milk, and every grain of your powder will contain the millionth of a grain of the medicinal substance. When the powder is of this strength, it is ready to employ in the further solutions and dilutions to be made use of in practice.

    A grain of the powder is to be taken, a hundred drops of alcohol are to be poured on it, the vial is to be slowly turned for a few minutes, until the powder is dissolved, and two shakes are to be given to it. On this point I will quote Hahnemann's own words. "A long experience and multiplied observations upon the sick lead me within the last few years to prefer giving only two shakes to medicinal liquids, whereas I formerly used to give ten."

The process of dilution is carried on in the same way as the attenuation of the powder was done; each successive dilution with alcohol reducing the medicine to a hundredth part of the quantity of that which preceded it. In this way the dilution of the original millionth of a grain of medicine contained in the grain of powder operated on is carried successively to the billionth, trillionth, quadrillionth, quintillionth, and very often much higher fractional divisions. A dose of any of these medicines is a minute fraction of a drop, obtained by moistening with them one or more little globules of sugar, of which Hahnemann says it takes about two hundred to weigh a grain. [From D. Stalker and C. Glymour, eds., *Examining Holistic Medicine* (Amherst, NY: Prometheus Books, 1985).]

25. See his website: *www.digibio.com*
26. Consider a lump of dough, of length L. If the dough is kneaded by being stretched and folded, then the particles will move about in the dough; this is a typical fluid dynamic situation. What will happen to a particular particle? Suppose we know its position with as much accuracy as we can achieve to be X from the end of the dough and suppose that the dough is stretched to twice its length and then folded. The distance of the particle from the end of the dough will double with each stretching and remain the same with each folding, i.e., it will become $2X$. This will continue until the particle crosses the half-way mark; then the distance after the fold will become $L - X$. The successive positions of the particle can be easily worked out. Suppose $L = 1.0$ and $X = 0.170654612$. Then the successive positions of the particle will be the series: 0.170654612; 0.341309224; 0.682618449; 0.634763103; 0.730473794; 0.539052412; 0.921895176; 0.156209648; 0.312419295; 0.499354362; 0.998708725; 0.002582550; 0.005165100; 0.010330200; 0.020660400; 0.041320800; 0.082641601; 0.165283203; 0.330566406; 0.661132812; 0.677734375; 0.64453125; 0.7109375; 0.578125; 0.84375; 0.3125; 0.625; 0.75; 0.5; 1; 0. What seems to happen is that the point moves to one end of the dough. Because we started with a "random" point, this implies that all points will end up in one place! This is obviously impossible. What is really happening is that information is being lost as the series develops: the initial precision is quite large; gradually this decreases. This is because we are working with a computer. As the stretching and folding proceeds, digits are shifted from the left to the right and the computer has not got any further information to fill in from the right, because it cannot know any more about the point's position than we have told it to start with. The transformation ceases when the point moves to the end of the dough. But in the real world we can go on folding and stretching indefinitely, so more and more information is found to replace the "lost" information that goes into producing chaos.
27. C. Bohler, P. E. Nielsen, and L. E. Orgel, "Template Switching Between PNA and RNA Oligonucleotides," *Nature* 376 (1995): 578–581.
28. I am indebted to Jack Cohen for this example.
29. In a computer a recursive function can be evaluated by using a stack while iteration can be done without.

CHAPTER 10

1. I. Kant, *Critique of Pure Reason* (London: Everyman, 1993).
2. Interestingly, Kant did not think that the truths of mathematics are analytic, because he thought that there was nothing about the ideas of twoness and addition that contained the idea of fourness. But this is really just a matter of names. There may be nothing in the word *two* that, when put together with another word *two* and the word *and* automatically gives the name *four*. However, this is equally true of an analytic definition, say, that a horse is a large equine quadruped. Analytic definition is not about the letters or other symbols used in words, or the ideas they may be associated with in our mind; it is about the assignation of meaning.
3. P. Davies, *The Mind of God* (New York: Simon and Schuster, 1992)
4. Imaginary numbers are ordinary numbers multiplied by the square root of –1.
5. Quaternions are numbers with four parts—one real and three imaginary. They were discovered (or invented if you prefer) by Sir William Rowan Hamilton.
6. A. N. Whitehead and B. A. W. Russell, *Principia Mathematica* (Cambridge: Cambridge University Press, 1910).
7. K. Gödel, "On Formally Undecidable Propositions of Principia Mathematica and Related Systems," first published in *Monatshefte für Mathematik und Physik* 38 (1931): 173–198.
8. For a discussion of Gödel's undecidability theorem, see R. Penrose, *The Emperor's New Mind* (Oxford: Oxford University Press, 1989), chapter 4. A more technical account can be found in E. Nagel, *Gödel's Proof* (London: Routledge and Kegan Pul, 1959).
9. G. Orwell, *Nineteen Eighty-Four* (London: Secker and Warburg, 1949).
10. All this is informal, but it can be formalized and has been. Church, in introducing his functional or lambda calculus, incidentally defined the operation of counting. If f is a function and x is a variable, then

    $$1 = \lambda f\,[f(x)]$$

    $$2 = \lambda f\,[f(f(x))]$$

    $$3 = \lambda f\,[f(f(f(x)))]$$

    etc., where the numbers 1, 2, 3, etc., are the result of the counting function f applied the appropriate number of times, i.e., f is producing a number corresponding to collections of objects.
11. G. Polya, *How to Solve It* (Princeton: Princeton University Press, 1945).
12. A simple argument goes like this: this process can be done by a partial evaluator program; a partial evaluator program will run on a computer; any program that can be run on a computer can be carried out by a human being; therefore, it can be done by a human being. I am not going to make such an argument, but it is how algebra *could* be done iteratively.

13. A prime number is one that has no factors except itself and 1. So 2 is a prime (the only even one) and so are 3, 5, 7, 11, and so on.
14. This can be shown by an argument along the following lines. Suppose we can devise a system of representation that uses every particle in the universe, *n*, and every discrete position in the universe, *p* (if *p* is not countable, then the *p* positions become indistinguishable). Then *L* is a number in the base *n*, having *p* digits. We cannot in fact know what this number is, because we don't actually know how many particles there are or how many discrete positions can be found for them to occupy. But we can safely say that, in a finite universe, *L* can be defined. What about $L + 1$? Does this number exist? The positivist would say "No," because the number cannot be produced as the result of the operation of a computer. It cannot be computed, because doing so would require a computer whose registers would be larger than the universe can hold.
15. R. Penrose, *The Emperor's New Mind* (Oxford: Oxford University Press, 1989), p. 116.
16. J. D. Barrow, *Pi in the Sky* (Oxford: Oxford University Press, 1992).
17. G. Mar and P. Grim, "Pattern and Chaos: New Images in the Semantics of Paradox," *Nous* 25 (1991): 659–693.
18. This means that truth can be partial—something can be one-third true, for example (for the moment we will not examine the ramifications of this assumption).
19. See chapter 4 for the explanation of this term.
20. This consists of the following two propositions, X and Y:
X. This proposition is true to the same extent as proposition Y is true.
Y. This proposition is true to the extent that proposition X is false.
This gives rise to the two equations

$$x_{n+1} - 1 - \mathrm{abs}(y_n - x_n)$$

$$y_{n+1} = 1 - \mathrm{abs}[(1 - x_n) - y_n]$$

where the subscripts represent the successive values of *x* and *y* as they are iterated. When these iterated values of *x* and *y* are computed and values of the complex number determined by *x* and *y* for which the number "escapes" from the diagram (i.e., those complex numbers are plotted whose modulus remains smaller than a certain fixed value), the resulting shape looks like that of a fantastic plant.
21. W. Whitman, "Song of Myself" in *Leaves of Grass* ([reprint] New York: HarperCollins, 2000).

CHAPTER 11

1. Compare: "Good and evil is the sickness of the mind": Zen Buddhist saying.
2. A. M. Turing, "On Computable Numbers, with an Application to the Entschei-

dungsproblem," *Proceedings of the London Mathematical Society, series 2* 42 (1936): 230–265.

3. To *grasp* the recursion we must perform the operation on a recursive system—that is why our consciousness is also connected with recursion. A recursive definition is only grasped by understanding the operation that it defines; in other words, recursion involves semantics (i.e., consciousness), whereas iteration does not.
4. G. Spencer-Brown, *Laws of Form* (London: George Allen & Unwin, 1979).
5. The most obvious difference is that the recursive definition requires the use of a stack, or push-down store, while the iterative process does not.
6. When we decide to actually perform the recursion we have to give it a parameter that tells the program how far the recursion will go; the stack is required to hold partial results.
7. H. R. Maturana and F. J. Varela, *Autopoiesis and Cognition: The Realization of the Living* (Dordrecht: Reidel Books, 1980).
8. T. Hardy, *New Review* (January 1890).
9. Mystical traditions have occurred also in Christianity and in Islam, although they were often at odds with the practices and beliefs of the majority.
10. P. Whigham, *The Poems of Catullus* (Harmondsworth, England: Penguin Classics 1966).
11. P. Totterdell, R. B. Briner, B. Parkinson, and S. Reynolds, "Fingerprinting Time Series: Dynamic Patterns in Self-Report and Performance Measures Uncovered by a Graphical Nonlinear Method," *British Journal of Psychology* 87 (1996): 43–60.
12. The logistic equation models the way in which food supply changes the population of an idealized species. $x' = A.x.(1 - x)$, where x is the population now, x' is the predicted population, and A is an amplification factor representing the rate of breeding. For $A > 3.72$, the series of values of x' is chaotic.
13. S. Rinaldi, "Laura and Petrarch: An Intriguing Case of Cyclical Love Dynamics," *SIAM Journal on Applied Mathematics*, 58 (1998): 1205–1221.
14. H-P. Roché, *Jules and Jim* (London: Calder, 1963), trans. Patrick Evans.

CHAPTER 12

1. Acceptance of contradiction—indeed recognition of its necessity as a creative force.
2. T. S. Kuhn, *The Structure of Scientific Revolutions* (Chicago: University of Chicago Press, 1970).
3. G. Spencer-Brown, *Laws of Form* (London: George Allen & Unwin, 1979).
4. R. Robertson, "The Uroboros," SCTPLS Conference, July 25, 1999, Clark Kerr Campus, University of California, Berkeley.
5. A similar theme was explored in the nineteenth century in E. A. Abbott, *Flatland: A Romance of Many Dimensions by A Square* ([reprint] New York: Dover Thrift Editions, 1992).

6. See, e.g., Y-P. Gunji, "Dynamically Changing Interface as a Model of Measurement in Complex Systems," *Physica D* 101 (1997), 27–54.
7. Gunji's use of the term *recursive* applies to a theoretical description of the problem (i.e., to an outside observer), whereas for the system itself, living or nonliving, the problem appears as an iterative one.
8. F. W. Lawvere, "Category Theory, Homology Theory and Their Applications II," *Lecture Notes in Mathematics* 92 (1969, whole issue).
9. The Turing *entscheidungsproblem* is a problem arising out of the Turing machine, which was already encountered in chapter 10. The question it poses is: can we predict, for a given Turing machine, whether it will ever halt (and so arrive at an answer represented by its end state)? This is the counterpart of the Hilbert problem posed in 1900 as to whether a proposition has a proof within a given symbolic logic system.
10. The fixed-point theorems deal with mapping of points onto themselves for a given function: a point is fixed if it is mapped onto itself by the function. Some points are fixed, e.g., $x = 0$ maps onto itself for the function $f(x) = x^2$; some are oscillating [e.g., $x = 1$ for $f(x) = (1 - x)$; and some are chaotic [e.g., $0 < x < 1$ for $f(x) = 3.8 \times (1 - x)$].
11. L. Wittgenstein, *Tractatus Logico-Philosophicus*, 6.4311 (London: Routledge & Kegan Paul, 1955).
12. Title essay in B. Russell, *Mysticism and Logic, and Other Essays* (London: Penguin, 1953).
13. A. Newberg, E. D'Aquili, and V. Rause, *Why God Won't Go Away: Brain Science and the Biology of Belief* (New York: Ballantine, 2001).
14. M. Shermer, "Is God All in the Mind?" *Science* 293 (2001): 54.
15. C. W. Perky, "An Experimental Study of Imagination," *American Journal of Psychology* 21 (1910): 422–452.
16. J. Segal and V. Fusella, "Effects of Imaged Pictures and Sounds on Detection of Auditory and Visual Signals," *Journal of Experimental Psychology* 83 (1970): 458–464.
17. R. J. Bird, Ph. D. Thesis, 1997, University of Newcastle, England.
18. It has been shown that children can distinguish their mother's face from that of a stranger as early as they can be tested-just a minute or so after birth, even before the umbilical cord has been cut; T. M. Field, D. Cohen, R. Garcia, and R. Greenberg, "Mother–Stranger Face Discrimination by the Newborn," *Infant Behavior and Development* 7 (1984): 19–25.
19. N. Chomsky, *Language and Problems of Knowledge: The Managua Lectures* (Cambridge: MIT Press 1988).
20. Leucippus, a contemporary of Democritus, also said the same thing, but not apparently in writing.
21. G. Leibniz, *Discourses on Metaphysics and the Monadology* ([reprint] Amherst, NY: Prometheus Books, 1992).
22. This is rather like the situation with the so-called fundamental particles, such as photons, which behave sometimes as discrete particles and sometimes as continuous waves. This is puzzling if we think that light must be one thing or another;

but if we imagine particle properties and wave properties both being contained within electromagnetic radiation, then it is not.

23. R. J. Bird, "Atoms, Molecules, and Emergence," *Acta Polytechnica Scandinavica* (Ma 91, 1998): 44–48.
24. K. Töllner, R. Popovitz-Biro, M. Lahav, and D. Milstein, "Impact of Molecular Order in Langmuir–Blodgett Films on Catalysis," *Science* 278 (1997): 2100–2102.
25. L. Sirovich and S. Karlsson, "Turbulent Drag Reduction by Passive Methods," *Nature* 388 (1997): 753–755.
26. It is worth remembering that modern physics has acknowledged the existence of objects a great deal stranger than these.
27. D. Bouwmeester et al., "Experimental Quantum Teleportation," *Nature* 390 (1997): 575–579.
28. J. Grinberg-Zylberbaum, M. Delaflor, L. Attie, and A. Goswami, "The Einstein–Podolsky–Rosen Paradox in the Brain: The Transferred Potential," *Physics Essays* 7 (1994): 422–427.
29. See C. G. Jung, *Synchronicity* (trans. by R. F. C. Hull) (Princeton, NJ: Princeton University Press, 1973), p. 15.
30. C. Flammarion, *The Unknown* (New York: Harper & Brothers, 1900).
31. Occam said various things: "It is vain to do with more what can be done with fewer" (Frustra fit per plura quod potest fieri per pauciora); "Complexity should not be introduced unnecessarily" (Pluralitas non est ponenda sine neccesitate). Something nearer to the usual version, "Entia non sunt multiplicanda praeter necessitatem," was written by a later scholar.
32. For instance, in pattern-classifying programs, it is usual to train a program on a set of patterns that belong to known classes. Classifiers that correctly identify all the patterns in the training set are then tested on new patterns. According to the Occam principle, the simpler the classifier, the more predictive power it is likely to have on new objects outside the training set. But this has been found not to be true. See P. M. Murphy and M. J. Pazzani, "Exploring the Decision Forest: An Empirical Investigation of Occam's Razor in Decision Tree Induction," *Journal of Artificial Intelligence Research* 1 (1994): 257–275; G. I. Webb, "Further Experimental Evidence Against the Utility of Occam's Razor," *Journal of Artificial Intelligence Research* 4 (1996): 397–417.
33. G. B. West, J. H. Brown, and B. J. Enquist, "A General Model for Ontogenetic Growth," *Nature* 413 (2001): 628–631.
34. One of the best programs for exploring fractals is Fractint: *http://www.fractint.org*
35. For example, the New England Complexity Studies Institute and the Santa Fe Institute in New Mexico.
36. J. Cohen and I. Stewart, *The Collapse of Chaos* (New York, London: Penguin, 1995).
37. J. Searle, *Minds, Brains and Science* (London: Penguin, 1992).
38. A. Huxley, *The Perennial Philosophy* (London: Chatto and Windus, 1969).
39. Plato, "The Simile of the Cave," Book 7 in *The Republic* (Harmondsworth, Middlesex: Penguin Classics, 1955).

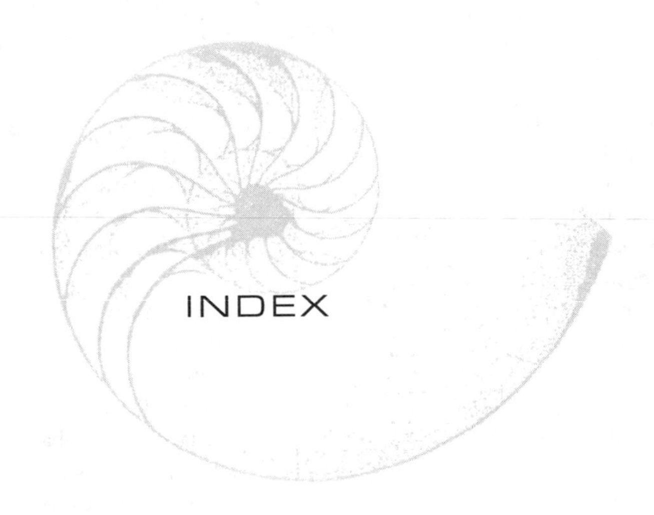

INDEX